W9-BYU-441

How to Run a Lathe

The Care and Operation
of
A Screw Cutting Lathe

By

J. J. O'Brien—M. W. O'Brien

42nd EDITION

Copyright 1942, by the
SOUTH BEND LATHE WORKS
All Rights Reserved

Martino Publishing
Mansfield Centre, CT
2013

Martino Publishing
P.O. Box 373,
Mansfield Centre, CT 06250 USA

ISBN 978-1-61427-474-2

© *2013 Martino Publishing*

All rights reserved. No new contribution to this publication may be reproduced, stored in a retrieval system, or transmitted, in any form or by any means, electronic, mechanical, photocopying, recording, or otherwise, without the prior permission of the Publisher.

Cover design by T. Matarazzo

Printed in the United States of America On 100% Acid-Free Paper

How to Run a Lathe

The Care and Operation
of
A Screw Cutting Lathe

By

J. J. O'Brien—M. W. O'Brien

42nd EDITION

Copyright 1942, by the
SOUTH BEND LATHE WORKS
All Rights Reserved

Also Copyright
1914, 1916, 1917, 1919, 1920, 1921, 1922, 1924, 1926, 1928, 1929, 1930, 1932, 1934, 1935, 1937, 1938, 1939, 1940, 1941
By South Bend Lathe Works

Reg. U. S. Pat. Off. &
Foreign Countries

Price 25 Cents, Postpaid
Leatherette Bound, Price 75 Cents, Postpaid
Stamps or Money Order of Any Country Accepted

SOUTH BEND LATHE WORKS
SOUTH BEND, INDIANA, U. S. A.

Printed in U. S. A.

CONTENTS

Chapter   Page

I. **HISTORY AND DEVELOPMENT OF THE LATHE**............... 3
The Tree Lathe; Early French Lathe; Maudslay Lathe; Modern Bench Lathe; Modern Standard Change Gear Lathe; Modern Quick Change Gear Lathe; Tool Room Lathe; Lathe Drives; The Size of the Lathe; Types of Lathes for Various Classes of Work; Features the Lathe Should Have.

II. **SETTING UP AND LEVELING THE LATHE**...................... 15
Setting Up and Leveling the Lathe; Lacing and Splicing Belts; Shifting Belts; Adjusting Belt Tension; Oiling the Lathe.

III. **OPERATION OF THE LATHE**..................................... 21
Principal Parts of a Lathe; Operation of Headstock; Spindle Speeds; Operation of Carriage and Apron; Operation of Tailstock; Power Carriage Feeds; Notes on Lathe Work.

IV. **LATHE TOOLS AND THEIR APPLICATION**......................... 27
Types of Lathe Tools; Position of Lathe Tool; Grinding Lathe Tool Cutter Bits; Cutting Power of Lathes.

V. **HOW TO TAKE ACCURATE MEASUREMENTS**...................... 37
Steel Scale; Outside Calipers; Inside Calipers; Hermaphrodite Calipers; Micrometers; Accuracy of a Lathe.

VI. **PLAIN TURNING (WORK BETWEEN CENTERS)**................... 43
Locating the Center; Drilling Center Hole; Lathe Dogs; Inserting and Removing Lathe Centers; Checking Alignment of Centers; Mounting Work Between Centers; Cutting Speeds; Facing; Turning; Machining to a Shoulder

VII. **CHUCK WORK** 53
Independent Chuck; Universal Chuck; Centering Work in the Chuck; Removing Chuck from Spindle; Hollow Spindle Chuck; Drill Chuck; Draw-in Collet Chuck, Handwheel and Hand-lever Types.

VIII. **TAPER TURNING AND BORING**................................. 59
Taper Turning with Compound Rest; Taper Boring with Compound Rest; Taper Turning with Tailstock Set Over; Taper Turning with Taper Attachment; Plain and Telescopic Type Taper Attachments; Morse Standard Tapers.

IX. **DRILLING, REAMING, AND TAPPING**............................ 65
Using Lathe as a Drill Press; Drill Pad; Crotch Center; Drilling Work Held in Chuck; Reaming; Tapping.

X. **CUTTING SCREW THREADS**...................................... 69
Standard Screw Thread Terms; Standard Change Gear Equipment; Quick Change Gear Equipment; Screw Thread Tables; Tools for Cutting Screw Threads; Use of Compound Rest; Thread Cutting Stop; Cutting Threads; Thread Dial; Screw Thread Forms; Metric Screw Threads.

XI. **SPECIAL CLASSES OF WORK**.................................... 87
Knurling; Face Plate Work; Filing and Polishing; Lapping; Machining Work on a Mandrel; Spring Winding; Coil Winding; Boring on Lathe Carriage; Use of Center Rest; Use of Follower Rest; Manufacturing Operations; Milling in the Lathe, etc.

PREFACE

ONE of the great needs of industry today is well trained workmen who are skilled with their hands and also trained to think about their work, diagnose troubles and suggest improvements. No man can hope to succeed in any line of work unless he is willing to study it and increase his own ability.

It is the purpose of this book to aid the apprentice in the machine shop and the student in the school shop to secure a better understanding of the fundamentals of the operation of a modern Screw Cutting Engine Lathe. In illustrating and describing the fundamental operations of modern lathe practice, we have made an effort to show only the best and most practical methods of machine shop practice in use in modern industries.

We are indebted to so many manufacturers, engineers, authors, educators, mechanics, and friends for assistance in the preparation of this book that it would be impossible to give them individual mention. However, we wish to express our appreciation for the cooperation that has made this work possible.

This is the 42nd edition of the book "How to Run a Lathe". The first edition was printed in 1907. Each succeeding edition has been revised and improved, and over 2,000 000 copies have been printed in the various languages.

"How to Run a Lathe"
in Spanish, Portuguese, and French

The book "How to Run a Lathe" is published in the English, Spanish, Portuguese, and French languages. See page 121.

See page 121.

SOUTH BEND LATHE WORKS.

Chapter I

HISTORY AND DEVELOPMENT OF THE
SCREW CUTTING LATHE

The screw cutting engine lathe is the oldest and most important of machine tools and from it all other machine tools have been developed. It was the lathe that made possible the building of the steamboat, the locomotive, the electric motor, the automobile and all kinds of machinery used in industry. Without the lathe our great industrial progress of the last century would have been impossible.

Early Screw Cutting Lathe

One of the earliest types of turning lathes was the tree lathe, shown in Fig. 1. A rope attached to a flexible branch overhead was passed around the work to revolve it. Later a strip of wood or "lath" was used to support the rope, and this is probably why the turning machine came to be known as a "lathe."

Fig. 1. Early Tree Lathe

One of the earliest screw cutting lathes that we have record of was built in France about 1740. Fig. 2 shows this lathe as it was illustrated in a book published in 1741. A hand crank was attached direct to the headstock spindle. The spindle of this lathe was geared to the lead screw, but there was no provision for changing the gears for cutting various pitches of screw threads.

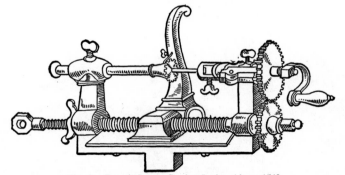

Fig. 2. French Screw Cutting Lathe, About 1740

(From Roe's ENGLISH AND AMERICAN TOOL BUILDERS; by permission of the publishers, McGraw-Hill Book Company, Inc.)

Fig. 3. Screw Cutting Lathe Made by Henry Maudslay, About 1797
(Courtesy Joseph Wickham Roe, Author of "English and American Tool Builders")

Henry Maudslay

Henry Maudslay, an Englishman, gave us the fundamental principles of the screw cutting engine lathe in a small lathe which he designed and built about 1797. On this lathe the gears used to connect the spindle with the lead screw could be changed, permitting the use of different gear ratios for cutting various pitches of screw threads.

Early American Lathes

Lathes were built in the United States between 1800 and 1830 with wood beds and iron ways. In 1836 Putnam of Fitchburg, Massachusetts built a small lathe with a lead screw. In 1850 iron bed lathes were made in New Haven, Connecticut, and in 1853 Freeland in New York City built a lathe, estimated 20 in. swing x 12 ft. bed, with iron bed and back geared head.

The Modern Bench Lathe

A modern geared screw feed bench lathe is shown in Fig. 4. This lathe has change gears to connect the spindle with the lead screw for cutting various pitches of screw threads and for power turning feeds, but does not have power cross feeds.

Fig. 4. A Modern Geared Screw Feed Bench Lathe

Fig. 5.　A Modern Standard Change Gear Lathe

The Modern Standard Change Gear Lathe

A modern standard change gear lathe is shown above in Fig. 5. This lathe is back-geared and has a four step cone pulley, providing eight spindle speeds. The lathe can be arranged for countershaft drive, or for direct motor drive.

The standard change gear lathe has a set of independent change gears for connecting the headstock spindle with the lead screw, as shown in Fig. 6. These gears may be arranged so that practically any pitch of screw thread may be cut. Change gears are also used for obtaining a wide range of power cross feeds and power longitudinal feeds for turning and facing operations.

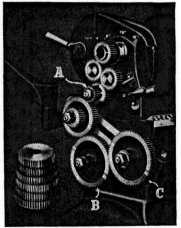

The standard change gear type of lathe is popular in the small shop, as it is less expensive than the quick change gear type of lathe. It is also widely used in industrial plants for production operations where few changes of threads and feeds are necessary. For this class of work the standard change gear lathe has an advantage in that when set up with the correct feeds for an operation the adjustments are not as easily tampered with and changed as they are on the quick change gear lathe.

Fig. 6.　End View of Standard Change Gear Lathe

Fig. 7. A Modern Quick Change Gear Lathe

Quick Change Gear Lathe

A quick change gear lathe is one in which the gearing between the spindle and lead screw is so arranged that changes for obtaining various pitches of screw threads may be made through a quick change gear box without removing or replacing a gear.

Fig. 7 shows a modern quick change gear lathe. The quick change gear mechanism attached to the left end of the lathe provides a series of 48 changes for cutting screw threads from 4 to 224 per inch, also a wide range of power feeds for turning, boring, and facing.

The quick change gear type of lathe is popular in busy shops where frequent changes of threads and feeds must be made such as in tool and die work, general repair and maintenance, and for some production operations.

Quick Change Gear Box

The interior of the quick change gear box is shown in Fig. 8. The gears in this gear box are shifted by levers operated from the front of the lathe and replace the independent change gears used on the standard change gear type of lathe.

Fig. 8. Interior of Quick Change Gear Box

(Patented)

Fig. 9. A Modern Underneath Motor Driven Toolroom Lathe

Toolroom Lathe

The Toolroom Lathe is the most modern type of back-geared screw cutting lathe and may be had with underneath motor drive, as shown above in Fig. 9, and also with pedestal motor drive and countershaft drive. Toolroom Lathes are usually given special accuracy tests during the process of manufacture and are equipped with taper attachment, thread dial indicator, draw-in collet chuck attachment, chip pan, and micrometer carriage stop.

The Toolroom Precision Lathe, as its name implies, is used in the toolrooms of industrial plants for making fine tools, test gauges and thread gauges, fixtures, jigs, etc., for making and testing the products manufactured.

A typical toolroom job, the making of a set of master thread gauges, is shown at right in Fig. 10. A threaded plug gauge for checking internal threads is being finished in the lathe and the round object in the lower left-hand corner is the threaded ring gauge for checking external threads.

Fig. 10. A Typical Toolroom Job

(Patented)

Fig. 11. A 13-in. Swing Lathe with Pedestal Motor Drive

Lathe Drives

Two types of drives for lathes are now in common use, the direct or individual motor drive, and the countershaft or group drive. (See page 18.)

Pedestal Motor Drive

The pedestal motor drive, shown above in Fig. 11, is one of the most practical types of direct motor drives for a lathe. An end view of the pedestal motor drive is shown in Fig. 12 at the right. The motor and countershaft are mounted on the pedestal back of the lathe. Power is transmitted from the motor to the countershaft by V-belts, and from the countershaft to the lathe spindle by a flat leather belt.

Belt tension adjustments "A" and "D" are provided for the cone pulley belt and the motor belts. A belt tension release lever "B" permits easy shifting of the cone pulley belt and also equalizes the pull of the belt between the pedestal and the lathe.

Fig. 12. End View of Pedestal Motor Drive

The Underneath Belt Motor Drive

The modern underneath belt motor drive shown in Figs. 13 and 14 is an efficient and practical direct drive equipment for a back-geared screw cutting lathe.   This drive is unusually compact and is silent, powerful and economical in operation.

The motor and driving mechanism are fully enclosed in the cabinet leg underneath the lathe headstock.   There are no exposed pulleys, belts or gears and no overhead belts or pulleys to obstruct vision or cast shadows upon the work.

Power is transmitted from the motor to the countershaft by V-belt and from the countershaft up through the lathe bed to the headstock cone pulley by a flat leather belt.

Adjustments "B" and "C" are provided for taking up belt stretch and for obtaining any desired tension on the motor belt and cone pulley belt. A belt tension release lever "A" conveniently located on the front of the cabinet leg permits easy shifting of the cone pulley belt. A hinged cover encloses the headstock cone pulley when the lathe is in operation. See page 19.

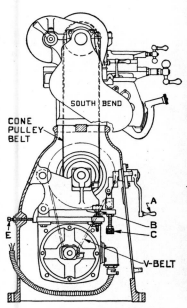

Fig. 13. Cross Section End View of Underneath Motor Drive

Fig. 14. Phantom View of Underneath Belt Motor Drive

Fig. 15.   A 9-in. Swing Bench Lathe with Adjustable Horizontal Motor Drive

Horizontal Motor Drive for Bench Lathes

The illustration above shows a 9-in. swing bench lathe equipped with an adjustable type horizontal motor drive unit. This is one of the most practical types of direct motor drive for a bench lathe.

The construction of the drive is shown below in Fig. 16. Belt tension adjustments "A" and "B" are provided for adjusting the tension of the cone pulley belt and the motor belt. A belt tension release lever "C" permits releasing the cone pulley belt tension so that the belt may easily be shifted from one step of the cone pulley to another. A flat leather belt is usually used between the cone pulleys and a V-belt is used between the motor pulley and the countershaft pulley.

Fig. 16.   End View of a Bench Lathe with Adjustable
Horizontal Motor Drive

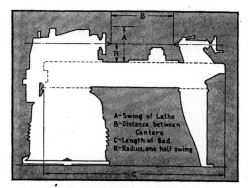

Fig. 17.   Size and Capacity of a Lathe

Size and Capacity of the Lathe

In the United States the size of a Screw Cutting Lathe is designated by the swing over bed and the length of bed, as indicated in Fig. 17 above. For example, a 16 in. x 8 ft. lathe is one having a swing over the bed "A" sufficient to take work up to 16 in. in diameter and having a bed length "C" of eight feet.

European tool manufacturers designate the size of a lathe by its radius "R" or center height.   For example, an 8-in. center lathe is a lathe having a radius of eight inches.   What the European terms an 8-in. center lathe, the American calls a 16-in. swing lathe.

The swing over the tool rest of the lathe is less than the swing over the bed, and the maximum distance between centers "B" is less than the length of the bed.   These figures must be considered carefully as they determine the size of work that can be machined between centers.

Selecting a Lathe for the Shop

When selecting a lathe, the most important point to consider is the size of the work.   The lathe should be large enough to accommodate the various classes of work that will be handled.   This is determined by the greatest diameter and length of work that will be machined in the lathe.   The lathe selected should have a swing capacity and distance between centers at least 10% greater than the largest job that will be handled.

Types of Lathes for Various Classes of Work

If the lathe you require is a large one, 13-in. swing or more, the floor leg type is recommended. If the lathe needed is of 9-in. or 10-in. swing, either a bench lathe or a floor leg lathe may be selected.   Floor leg lathes are usually more rigid than a lathe mounted on a bench because the heavy cast iron legs provide a sturdy, heavy support.   If a bench lathe is used, the bench should be sturdy and rigid and should have a top of 2-in. lumber.

Type of Drive for the Lathe

The overhead countershaft drive is used principally in factories where a number of countershaft driven machines are operated from a single lineshaft. This method is called "group drive" and has advantages when most of the shop machinery will be in operation at the same time. (See page 18.)

In some shops the individual motor drive is more practical and efficient than a lineshaft drive because a small motor can be used to drive each machine and the cost of installing hangers, lineshafting, etc., is eliminated. Also when the machine is not being used, the motor power may be shut off.

Change Gear Equipment

Quick Change Gear Lathes are preferred in busy shops where frequent changes of threads and feeds are required. Standard Change Gear Lathes are used in production shops on jobs that do not require many changes for threads or feeds, also in small shops that do not have a great deal of lathe work.

Fig. 18-A. Above. Close-up showing hinged guard for bull gear lock pin.

Fig. 18. Back-Geared Headstock, Gear Guards Removed

Features the Lathe Should Have

In considering a metal working lathe for the shop it is well to bear in mind that the lathe will be used for many classes of work and that if carefully selected it should give years of satisfactory service.

The Headstock

The headstock is the most important unit of the lathe and should be back-geared, as shown in Fig. 18. The back gears provide the slow spindle speeds and power required for taking heavy cuts on large diameter work. Modern lathes are equipped with back gears having a quick acting bull gear lock which permits engaging or disengaging the back gears without using a wrench.

Fig. 19. Hardened Alloy Steel Headstock Spindle with Superfinished Spindle
Bearing Surfaces

Headstock Spindle and Bearings

The lathe headstock spindle should be made of a good quality alloy spindle steel, and for best service should be heat treated after it is machined, and all bearing surfaces, including the taper hole, should be carburized, hardened and ground.

The Journal bearing surfaces on the spindle should be "superfinished" to a smoothness of five micro inches (.000005")*. When equipped with a superfinished spindle and proper oiling facilities, the lathe can be operated at the high speeds essential for the efficient use of modern tungsten-carbide tipped cutter bits without danger of overheating or scoring the spindle bearings.

Integral cast iron bearings provide the best support for the spindle and are the most durable. This construction is preferable to bronze bushings or anti-friction bearings because it permits using a large diameter spindle, which is essential for maximum strength and rigidity. The spindle diameter can be enlarged to fill the entire space that would otherwise be required for bushings or bearing cups.

Lathe Bed Construction

The lathe bed is the foundation on which the lathe is built, so it must be substantially constructed and scientifically designed. Fig. 20 shows an end view of a lathe bed, which is an example of modern design.

Prismatic V-ways have been found to be the most accurate and serviceable type of ways for lathe beds and have been adopted by most of the leading machine tool builders. The two outer V-ways (1 and 4) guide the lathe carriage, while the inner V-way and flat way (2 and 3) align headstock and tailstock.

The V-ways of the lathe bed are carefully hand scraped so that the headstock, carriage and tailstock are perfectly fitted and aligned parallel to the axis of spindle the entire length of bed.

*Measurements in microinches *rms.*

Fig. 20. End View of Lathe Bed Showing
Prismatic V-Ways

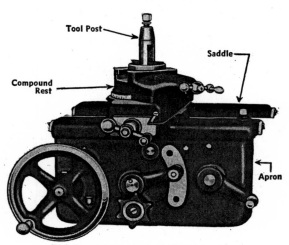

Tool Post

Saddle

Compound Rest

Apron

Fig. 21. A Well Designed Lathe Carriage

The Lathe Carriage

The lathe carriage includes the apron, saddle, compound rest and tool post. Since the carriage supports the cutting tool and controls its action, it is one of the most important units of the lathe. The carriage shown in Fig. 21 is modern and practical.

The Apron is of double wall construction with all gears made of steel. A powerful multiple disc clutch is provided for the automatic friction feeds. An automatic safety device prevents the half nuts and automatic feeds from being engaged at the same time.

The threads of the lead screw are used only for thread cutting. A spline in the lead screw drives a worm in the apron which operates the automatic power carriage feeds.

Interior of Apron

The interior of the apron is shown in Fig. 22 at right. The spline in the lead screw which drives the worm for operating the power longitudinal feeds and power cross feeds is clearly shown.

The half nuts for thread cutting are dovetailed into the back wall of the apron.

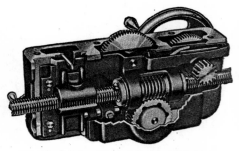

Fig. 22. Interior View of Double Wall Apron

PRECISION LEVELS

Fig. 23. Leveling the Lathe

Chapter II

SETTING UP AND LEVELING THE LATHE

A new lathe should be very carefully unpacked and installed so that all of the fine accuracy that has been built into the lathe by the manufacturer will be retained.

Do not allow a hammer or crow bar to strike the lathe while unpacking as this may cause serious damage. Look carefully in all packing material for small parts, instruction material, etc. Study all reference books and instruction sheets carefully before setting up the lathe.

Clean the new lathe thoroughly with a stiff brush and kerosene. Wipe with a clean cloth and then immediately cover all unpainted surfaces with a film of good machine oil to prevent rusting. Wipe off the old oil occasionally and do not allow dust, chips or dirt to accumulate. Cover the lathe with canvas when not in use. Keep the finished surfaces clean and well oiled and the lathe will retain its new appearance.

Solid Floor Required

It is very important that the lathe be set on a solid foundation and that it is carefully and accurately leveled. An erection plan showing how to set up and level the lathe is included in the shipment of the lathe. For best results the lathe should be set on a concrete foundation. A wood floor should be braced to prevent sagging and vibration if it is not substantially constructed.

The lathe may be leveled by placing shims of hard wood or metal under the legs, as shown in Fig. 23. If the lathe is not leveled it will not set evenly on all four legs and the weight of the lathe will cause the lathe bed to be twisted, throwing the headstock out of alignment with the V-ways of the bed and causing the lathe to turn and bore taper. If the lathe is not level it cannot turn out accurate work.

15

Use a Precision Level for Leveling the Lathe

Use a precision level that is at least twelve inches long and sufficiently sensitive to show a distinct movement of the bubble when a .003 in. shim is placed

Fig. 24. A Precision Level

under one end of the level. Level across the lathe bed at both the headstock end and the tailstock end, as shown in Fig. 23, page 15.

Bolt the Lathe to the Floor

Use lag screws or bolts to secure the lathe to the floor. If the lathe is set on a concrete floor or foundation, mark the location of the bolt holes and drill holes in the concrete with a star drill. Use expansion bolts or set bolts in melted lead or melted sulphur. Check the leveling of the lathe after bolting the lathe to the floor or bench.

Leveling Underneath Belt Motor Driven Lathes

When placing shims under the cabinet leg at the headstock end of the Underneath Motor Drive Lathe, use the shims only at the bolt pads. There should be clearance under the cabinet leg all the way round except at the two pads where the bolts go through the leg into the floor.

Bench Lathes

Bench Lathes should be mounted on a substantial bench, providing rigid support and leveled as outlined above. The bench top should be about 28 in. high and if made of wood should be of 2-in. lumber. The bench should be securely bolted to the floor so there will be no danger of the bench shifting and throwing the lathe out of level. Some bench lathes have leveling screws in the right leg which may be used for making the final leveling adjustments.

Readjust Shims

It may be necessary to readjust the shims under the lathe legs from time to time to compensate for settling of the building, even if the lathe is set on a concrete floor. For this reason the legs should not be bedded in concrete but should be bolted to the floor.

If at any time the lathe does not bore a straight hole, this is an indication that the lathe is no longer perfectly level and the shims should be readjusted.

Checking the Leveling of Lathe

After leveling the lathe, place a bar of steel one inch or larger in diameter in the chuck and machine two collars of equal diameter three or four inches apart, as shown in Fig. 25. Take a very light finishing cut across both of these collars without changing the adjustment of the cutter bit. Measure the diameter of each collar carefully with a micrometer.

If the collars are not the same diameter, this is an indication that the level used in setting up the lathe was not sufficiently sensitive. The leveling may be perfected by adjusting the shims under the front and back legs at the tailstock end of the lathe until the collars on the test piece are turned the same diameter.

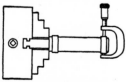

Fig. 25. Method of Testing Leveling of Lathe

Lacing Leather Belts

Leather belts may be joined by lacing with gut or rawhide thongs, as shown in Fig. 26. The smooth side of the belt should run next to the pulley, and the lacing should not be crossed on the pulley side.

Trim the ends of the belt square and to the length required. Use a steel tape and measure over the pulleys to determine the correct belt length.

Punch or drill holes just large enough for the lace in each end of belt, as shown in Fig. 26. (Wide belts require more holes.)

If a round gut lace is used, cut straight grooves $\frac{1}{8}$ in. wide $\frac{1}{16}$ in. deep from holes to end of belt on smooth side. This will make the belt run smoothly over the pulleys.

Start lace through center holes from grooved side and pull both ends through evenly. Work out both ways, then back to center. Be careful not to kink the lacing.

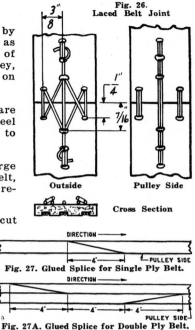

Fig. 26.
Laced Belt Joint

Outside   Pulley Side

Cross Section

DIRECTION ⟶

PULLEY SIDE

Fig. 27. Glued Splice for Single Ply Belt.

DIRECTION ⟶

PULLEY SIDE

Fig. 27A. Glued Splice for Double Ply Belt.

Do not cross lacing on grooved side of belt which runs next to the pulleys. Fasten ends of lace, as shown.

Glued Belt Splice

Many mechanics prefer a glued belt splice to a laced joint because the glued splice will usually run more smoothly. To make a good glued splice, both ends of the belt must be tapered uniformly, as shown in Fig. 27. Double ply belts should be split a short distance and both parts tapered, as shown in Fig. 27A.

Any good belt cement may be used. Experience has shown that best results are obtained with "airplane dope" or a good acetone cement. When cutting belts for a glued splice, be sure to allow a sufficient length for lapping the ends together.

Wire Belt Hooks

There are a number of good wire belt hooks on the market that can be used for fastening belt ends together. The belt hooks save time and are easy to use.

Wire belt hooks are often used for belts that are not shifted while the machine is in operation, but should never be used on belts that are shifted while the machine is running.

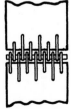

Fig. 28. Wire Belt Hooks

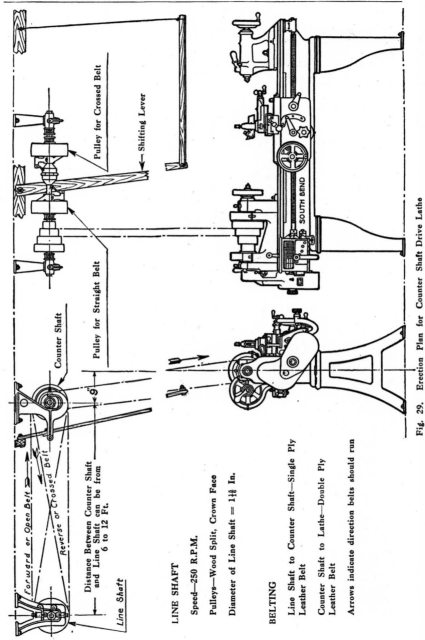

LINE SHAFT

Speed—250 R.P.M.

Pulleys—Wood Split, Crown Face

Diameter of Line Shaft = $1\frac{15}{16}$ In.

BELTING

Line Shaft to Counter Shaft—Single Ply
Leather Belt

Counter Shaft to Lathe—Double Ply
Leather Belt

Arrows indicate direction belts should run

Fig. 29.   Erection Plan for Counter Shaft Drive Lathe

Use Leather Belts

Good quality leather belts are best for lathe cone pulleys. They have sufficient elasticity to transmit power efficiently and they give good service. No belt dressing is required if the belts are kept clean and dry. Machine oil will cause belts to slip.

Shifting Belts

To shift the belt on a countershaft driven lathe the beginner should stop the lathe and shift by hand by pulling belt and slipping belt to the desired position.

Fig. 30 shows the method of shifting the belt on a countershaft driven lathe while the lathe is running. Using a stick, the operator pushes the belt from one cone pulley step to another.

To shift the belt on the countershaft cone to a larger step, use a long belt stick with a steel pin in the end. While the countershaft is revolving, give the belt a sharp push and twist with the pin on the end of the stick.

Fig. 30. Shifting the Belt of a Countershaft Drive Lathe

Underneath Motor Driven Lathes

Motor driven lathes must always be stopped before the cone pulley belt is shifted. The belt tension release lever "A" (Fig. 31) permits releasing the cone pulley belt tension for easy shifting of the belt to change spindle speeds.

Stop screw "E" eliminates play when lever "A" is in the lower position. This screw must be unscrewed several revolutions before any belt tension adjustments are made and must be readjusted afterward.

Screw "B" is for adjusting the tension of the motor V-belts. The adjustment is made by turning the adjusting nuts on this screw, above and below the motor support bracket.

Knurled knob "C" is for adjusting the tension of the cone pulley flat belt. This adjustment should be made with lever "A" in the lower position.

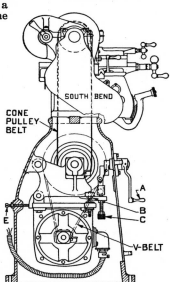

Fig. 31. Cross Section of Underneath Belt Motor Drive Showing Cone Pulley Belt and V-Belts

The belts should be just tight enough to transmit the required power without slipping. When it is properly adjusted, pressing the hand against the flat belt near the cone pulley should depress the belt about ½". Pressing against the V-belt midway between the pulleys should depress the belt about 1".

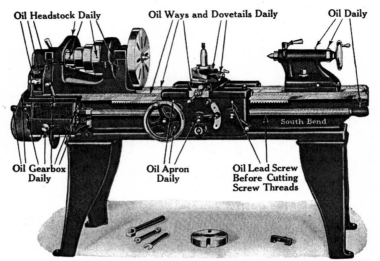

Oil Headstock Daily   Oil Ways and Dovetails Daily   Oil Daily

Oil Gearbox   Oil Apron   Oil Lead Screw
Daily   Daily   Before Cutting
                                                 Screw Threads

Fig. 32. Oiling Chart for a Lathe

Oiling the Lathe

Oil every bearing on the lathe before starting it as directed on the oiling chart which you will find packed with the lathe. Do not attempt to oil the lathe while it is running. Use a good grade of machine oil or S.A.E. No. 10 engine oil unless instructions supplied with lathe specify otherwise. Oil twice daily for the first week and once a day thereafter. Do not run the lathe spindle more than 500 R.P.M. until it has been thoroughly broken in.

Keeping the lathe well oiled has much to do with the life of the lathe and the quality of the work it will turn out. Follow these directions carefully if you wish to keep your lathe in first class condition.

Always oil in the same order so that no oil holes will be missed. If you do this the oiling will become a habit and will require only a very short time.

Do not use an excess of oil. A few drops in each oil hole is sufficient, and if more is applied it will only run out of the bearings and get on the lathe, making it necessary for you to clean it more frequently.

Oil the countershaft each time the lathe is oiled. If the lathe is motor driven, oil the motor bearings once a week as specified in instructions supplied with lathe.

After you have completed the process of oiling the lathe and countershaft, wipe off the excess oil around the bearings with a clean cloth or waste. Keep the lathe clean. Do not allow oil, dirt, chips or rust to collect any place on the lathe.

Chapter III

OPERATION OF THE LATHE

Before starting a new lathe, the operator should carefully study the action of the various parts and become thoroughly familiar with the operation of all control levers and knobs.

The principal parts of the lathe are shown below in Fig. 33. Become familiar with the name of each part as they will be referred to frequently in the following pages where detailed information on the operation of the lathe is given.

Do not operate the lathe under power until it is properly set up and leveled, as outlined on page 15. Also make sure that all bearings have been oiled and that the belt tension is correct. Always pull the cone pulley belt by hand to make sure the lathe runs free before starting the lathe under power.

Fig. 33. Names of the Principal Parts of a Lathe

Fig. 34. Operating Parts of Lathe Headstock

Operation of Headstock

Spindle speeds are changed by shifting the belt from one step of the cone pulley to another and by engaging or disengaging the back gears. The cone pulley steps are numbered in the illustration above to correspond with the numbers in the tabulation on page 23, which shows the normal spindle speeds for various sizes of lathes.

Direct Belt Drive

To arrange the lathe headstock for direct belt drive, push the back gear lever back as far as it will go; then pull out and up on the bull gear lock pin and revolve the cone pulley slowly by hand until the bull gear lock slides into position and locks the cone pulley to the spindle.

Back Geared Drive

To engage the back gears for slow spindle speeds, pull the bull gear lock pin out and push it down to disconnect the cone pulley from the spindle; then pull the back gear lever forward. Revolve the cone pulley by hand to make sure the back gears are properly engaged. Do not engage the back gears while the lathe spindle is revolving.

Bull Gear Lock, Plunger Type

On some lathe headstocks the plunger type bull gear lock is used. For direct belt drive on these lathes the bull gear lock pin is pushed in, and for back-geared drive it is pulled out.

Feed Reverse Lever

The feed reverse lever on the left end of the headstock has three positions; up, central and down. The central position is neutral, and when in this position all power carriage feeds are disconnected. When the lever is in either the "up" position or "down" position the power carriage feeds will be in operation.

Spindle Speeds of Lathes

The standard spindle speeds for various sizes of South Bend Lathes are listed in the tabulations below. The columns under which the speeds are listed are numbered 1, 2, 3, and 4 to correspond with the numbers on the cone pulley steps in Fig. 34, page 22. For example, those spindle speeds that are listed under Column 1 are obtained when the cone pulley belt is placed on the cone pulley step marked 1 in the illustration, Fig. 34.

STANDARD SPINDLE SPEEDS OF SOUTH BEND LATHES IN REVOLUTIONS PER MINUTE

Size of Lathe	Counter-shaft Speed	Spindle Speeds Direct Belt Drive				Spindle Speeds Back Gear Drive			
		1	2	3	4	1	2	3	4
9-inch	300	658	370	212	—	127	72	41	—
10-inch 11⁄16-in. Collet	300	700	434	277	—	129	79	50	—
13-inch Std. and Q. C.	369	875	567	373	239	128	81	54	34
14½-inch Std. and Q. C.	335	800	482	300	181	121	72	45	27
16-inch Std. and Q. C.	274	725	438	277	171	91	55	35	21

SPINDLE SPEEDS OF LATHES WITH TWO SPEED MOTOR DRIVE COUNTERSHAFT

Size of Lathe	Counter-shaft Speed	Spindle Speeds Direct Belt Drive			Spindle Speeds Back Gear Drive		
		1	2	3	1	2	3
9-inch	579	1270	716	408	*	*	*
	300	658	370	212	127	72	41
10-inch 1-in. Collet	579	1357	837	535	*	*	*
	300	700	434	277	129	79	50

*When using high countershaft speed, back gears should not be engaged.

High Spindle Speeds

When high spindle speeds are required, special drive equipment is used to provide higher speeds than the standard speeds shown in the tabulation above. More power is required for operating the lathe at high speeds than for operating the lathe at normal speeds. The spindle bearings must be well lubricated and should have ample clearance for oil film. A direct belt drive to the spindle is essential for smooth, vibration-free operation at high speed.

Com-
Pound
Rest
Knob

Carriage
Lock
Screw

Feed
Change
Lever

Cross
Feed
Knob

Half Nut
Lever

Apron
Hand
Wheel

Automatic
Feed Fric-
tion Clutch

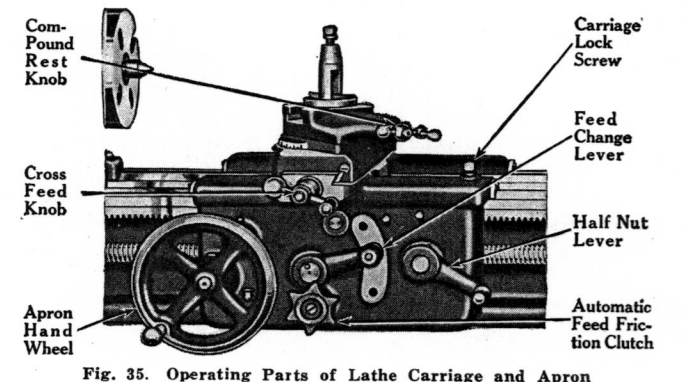

Fig. 35. Operating Parts of Lathe Carriage and Apron

Operation of Lathe Carriage and Apron

The principal operating parts of the lathe carriage and apron are shown above in Fig. 35. The apron hand wheel is turned to move the carriage along the lathe bed, and the cross feed knob and compound rest knob are turned to move the tool rest in and out. The carriage lock screw is used to lock the carriage to the lathe bed. This screw should never be tightened except for facing or cutting-off operations.

Fig. 35A. Micrometer Collar
on Cross Feed Screw

Micrometer Collars

Each graduation of the micrometer collars on the cross feed knob and compound rest knob represents a movement of the compound rest of one-thousandth of an inch. The graduated collars may be set at zero by releasing the set screws which lock them in position.

Power Carriage Feeds

The automatic feed friction clutch controls the operation of both the automatic power longitudinal feed and the automatic power cross feed. To engage the clutch, turn the club knob to the right; to disengage, turn to the left. The direction of the feed is controlled by the position of the reverse lever on the headstock. (See page 22.)

The feed change lever has three positions: "up" for longitudinal feeds, "down" for cross feeds, and "center" for neutral.

The half nut lever is used only for thread cutting. The feed change lever must be in the "center" or neutral position before the half nuts can be engaged.

Operation of Tailstock

The tailstock may be locked on the lathe bed at any position by tightening the clamp bolt nut. To lock the tailstock spindle, tighten the binding lever by pulling it forward.

Binding Lever

Clamp
Bolt
Nut

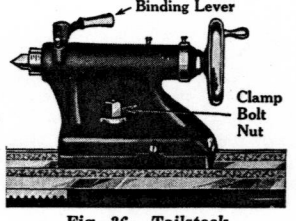

Fig. 36. Tailstock

Power Carriage Feeds on Quick Change Gear Lathes

A wide range of power longitudinal feeds and power cross feeds is available on all Quick Change Gear Lathes. To obtain any desired feed it is only necessary to arrange the levers on the gear box according to the direct reading index chart shown in Fig. 38. The threads per inch are shown in large figures on the index chart below. The smaller figures indicate the power longitudinal turning feeds in thousandths of an inch.

Fig. 37. Quick Change Gear Mechanism

16-INCH SOUTH BEND QUICK CHANGE GEAR LATHE										
SLIDING GEAR	TOP LEVER	THREADS PER INCH—FEEDS IN THOUSANDTHS								AUTOMATIC CROSS FEED EQUALS .375 TIMES LONGITUDINAL FEED
IN	LEFT	4 .0841	4½ .0748	5 .0673	5½ .0612	5¾ .0585	6 .0561	6½ .0518	7 .0481	
	CENTER	8 .0421	9 .0374	10 .0337	11 .0306	11½ .0293	12 .0280	13 .0259	14 .0240	
	RIGHT	16 .0210	18 .0187	20 .0168	22 .0153	23 .0146	24 .0140	26 .0129	28 .0120	
OUT	LEFT	32 .0105	36 .0093	40 .0084	44 .0076	46 .0073	48 .0070	52 .0065	56 .0060	
	CENTER	64 .0053	72 .0047	80 .0042	88 .0038	92 .0037	96 .0035	104 .0032	112 .0030	
	RIGHT	128 .0026	144 .0023	160 .0021	176 .0019	184 .0018	192 .0017	208 .0016	224 .0015	

Fig. 38. Index Chart for Quick Change Gear Lathe

Fig. 39. Standard Change Gear Mechanism

Power Carriage Feeds on Standard Change Gear Lathes

Standard Change Gear Lathes are equipped with a set of independent change gears for cutting screw threads and obtaining various power longitudinal feeds and power cross feeds. Compound gearing is used for fine threads and feeds. See page 114.

A large "screw gear" "C" should be placed on the lead screw and a small "stud gear" "A" on the reverse stud. These two gears should be connected with idler gears "B," as shown. To obtain finer or coarser feeds, use a smaller or larger "stud gear."

Notes on Lathe Work

A mixture of red lead and machine oil is a good lubricant for the tailstock center of a lathe.

A precision level that will show an error of .003 in. per foot should be used to level the lathe when installing, as a level lathe will assure precision accuracy of the work.

Clean and oil the threads before screwing a chuck or face plate onto the lathe spindle.

After grinding a tool, hone it to a keen edge with an oil stone— the cutting edge will last longer.

Always make sure the spindle tapers of the lathe are clean and free from burrs and dirt before inserting the lathe centers.

If the face plate or chuck does not run true, examine the shoulder of the lathe spindle and face of hub on face plate or chuck back for burrs, dirt, etc.

When cutting screw threads in steel, use a small brush to spread oil on the work preceding each cut. Lard oil is preferable, but a good machine oil or cutting oil will do.

Use Flat Leather Belts

Flat leather belts are recommended for use on lathe cone pulleys.

Leather belts are better than canvas or rubber belts for use on lathe cone pulleys. Leather belts are more efficient, last longer, have more elasticity and give better service.

If a belt has a tendency to come off from the pulley there is something wrong. Usually the pulleys are out of alignment. Find out what the trouble is and remedy it. Do not try to hold the belt on the pulley with a brace.

Notes on Belts and Pulleys

To find the approximate length of a belt, multiply half the sum of the pulley diameters by 3-1/7 and add twice the distance between the pulley centers.

The smooth side of the belt should always run next to the pulley.

Keep belts clean and dry. Do not allow moisture, machine oil or dirt to collect on them.

A pulley should be about 10% wider than the belt.

Driving pulleys for shifting belts should have flat face, all other pulleys should be crowned.

For stepped or flanged pulleys double ply belting is better than single ply belting.

Don't shift a moving belt by hand; use a stick or belt shifter.

Never put a belt on a pulley while it is revolving rapidly.

A belt may run crooked if the ends are not cut square before lacing, or if laced unevenly.

Don't run belts too tight, or with the flesh side next to the pulley.

Chapter IV

LATHE TOOLS AND THEIR APPLICATION

In order to machine metal accurately and efficiently, it is necessary to have the correct type of lathe tool with a keen, well supported cutting edge, ground for the particular kind of metal being machined, and set at the correct height.

High speed steel cutter bits mounted in forged steel holders, as shown in Figs. 40, 46, 48 and 50, are the most popular type of lathe tools.

The boring tool, cutting-off tool, threading tool and knurling tool are required for various classes of work that cannot be readily accomplished with the regular turning tool.

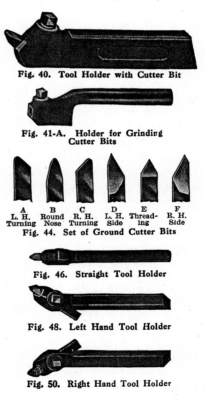

Fig. 40. Tool Holder with Cutter Bit

Fig. 41-A. Holder for Grinding Cutter Bits

Fig. 41. Unground Cutter Bit

Fig. 42. Cutter Bit After Grinding

Fig. 43. Boring Tool

A   B   C   D   E   F
L. H. Round R. H. L. H. Thread- R. H.
Turning Nose Turning Side ing Side
Fig. 44. Set of Ground Cutter Bits

Fig. 45. Cutting-off Tool

Fig. 46. Straight Tool Holder

Fig. 47. Formed Threading Tool

Fig. 48. Left Hand Tool Holder

Fig. 49. Knurling Tool

Fig. 50. Right Hand Tool Holder

Correct Height of Cutting Edge

The cutting edge of the cutter bit should be about 5° above center, or $\frac{3}{64}$ in. per inch in diameter of the work, as shown in Fig. 51 at right, for ordinary straight turning. The position of the cutter bit must be taken into consideration when grinding the various angles, as the height of the cutter bit determines the amount of front clearance necessary to permit free cutting.

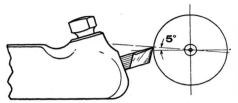

Fig. 51. Cutting Edge of Cutter Bit 5° Above Center for Straight Turning

The cutting edge of the cutter bit should always be placed exactly on center, as shown in Fig. 52, for all types of taper turning and boring, and for cutting screw threads, also for turning brass, copper and other tenacious metals.

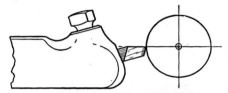

Fig. 52. Cutting Edge of Cutter Bit Exactly on Center for Thread Cutting, Taper Turning, Machining Brass, Copper, etc.

Tool Angle Varies with Texture of Work

The included angle of the cutting edge of a cutter bit is known as the tool angle or angle of keenness and varies with the texture of the work to be machined. For example, when turning soft steel a rather acute angle should be used, but for machining hard steel or cast iron the cutting edge must be well supported and therefore the angle is less acute.

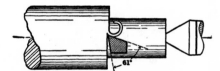

Fig. 53. Tool Angle for Machining Soft Steel

It has been found that an included angle of 61° is the most efficient tool angle for machining soft steel. This is the angle of the cutter bit as shown in Fig. 53.

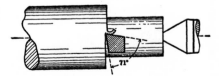

Fig. 54. Tool Angle for Machining Cast Iron

For machining ordinary cast iron, the included angle of the cutting edge should be approximately 71°, as shown in Fig. 54. However, for machining chilled iron or very hard grades of cast iron, the tool angle may be as great as 85°.

Grinding Lathe Tool Cutter Bits

The angle of the cutter bit with the bottom of the tool holder must be taken into consideration when grinding cutter bits.

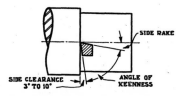

The side clearance (Fig. 55) is to permit the cutting edge to advance freely without the heel of the tool rubbing against the work.

Fig. 55. Correct Side Clearance and Side Rake of Cutter Bit

The front clearance (Fig. 56) is to permit the cutting edge to cut freely as the tool is fed to the work.

Too much clearance will weaken the cutting edge so that it will break; but insufficient clearance will prevent the tool from cutting.

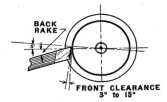

Side rake and back rake (Figs. 55 and 56) also facilitate free cutting. For cast iron, hard bronze and hard steel, very little side rake or back rake are required. (See page 28.)

Fig. 56. Correct Front Clearance and Back Rake of Cutter Bit

The angle of keenness (Fig. 55) may vary from 60° for soft steel to nearly 90° for cast iron, hard steel, bronze, etc.

Figs. 57 to 61, inclusive, show the various steps in grinding a cutter bit for general machine work. Honing the cutting edge (Fig. 62) will improve the quality of the finish and lengthen the life of the tool.

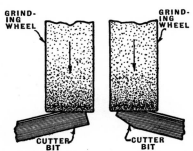

Fig. 57. Grinding Left Side of Cutter Bit

Fig. 58. Grinding Right Side of Cutter Bit

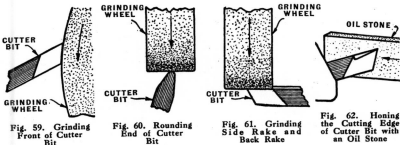

Fig. 59. Grinding Front of Cutter Bit

Fig. 60. Rounding End of Cutter Bit

Fig. 61. Grinding Side Rake and Back Rake

Fig. 62. Honing the Cutting Edge of Cutter Bit with an Oil Stone

Cutter Bit for Rough Turning

Figs. 63 and 64 illustrate an excellent tool for taking heavy roughing cuts to reduce the diameter of a steel shaft to the approximate size desired. This tool will cut freely but does not produce a very smooth finish. When using this type of tool it is advisable to leave sufficient stock for a finishing cut with the round nosed tool shown at the bottom of the page.

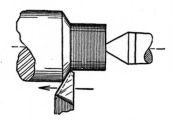

Fig. 63. Application of Roughing Tool

Grind the tool to the shape shown in Fig. 64, and see Figs. 55 and 56 on page 29 for information on grinding the correct front clearance, etc.

The cutting edge of the tool is straight and the point is only slightly rounded. A very small radius at the point (approximately $\frac{1}{64}$ in.) will prevent the point of the tool from breaking down but will not impair the free cutting quality of the tool.

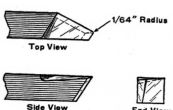

Fig. 64. Detail of Roughing Tool

The tool angle or included angle of the cutting edge of this tool should be approximately 61° for ordinary machine steel. If a harder grade of alloy or tool steel is to be machined, the angle may be increased, and if free cutting Bessemer screw stock is to be machined, the angle may be slightly less than 61°.

Hone the cutting edge of the tool with a small oil stone. This will lengthen the life of the tool and it will cut better.

Cutter Bit for Finish Turning

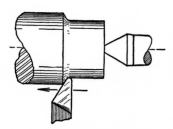

Fig. 65. Application of Finishing Tool

Figs. 65 and 66 illustrate a round nosed turning tool for taking finishing cuts. The tool is very much the same shape as the more pointed tool for rough turning shown above, except that the point of the tool is rounded. (Approximately $\frac{1}{32}$ in. to $\frac{1}{16}$ in. radius.)

This tool will produce a very smooth finish if, after grinding, the cutting edge is well honed with an oil stone and a fine automatic power carriage feed is used.

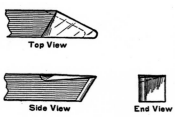

Fig. 66. Detail of Finishing Tool

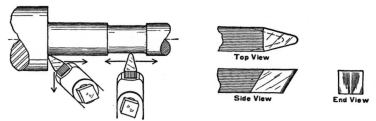

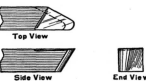

Fig. 67. Application of Round Nosed Tool   Fig. 68. Detail of Round Nosed Tool Bit

Round Nosed Turning Tool

The round nosed turning tool shown above is ground flat on top so that the tool may be fed in either direction, as indicated by the arrows in Fig. 67. This is a very convenient tool for reducing the diameter of a shaft in the center. The shape of the cutter bit is shown in Fig. 68, and the correct angle for the front clearance and side clearance can be obtained by referring to Figs. 55 and 56, page 29.

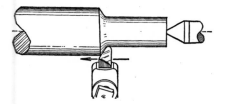

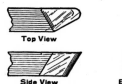

Fig. 69. Application of Right Hand Turning Tool   Fig. 70. Detail of Right Hand Turning Tool

Right Hand Turning Tool

The right hand turning tool shown above is the most common type of tool for general all around machine work. This tool is used for machining work from right to left, as indicated by the arrow in Fig. 69. The shape of the cutter bit is shown in Fig. 70. See page 29 for correct angles of clearance.

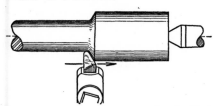

Fig. 71. Application of Left Hand Turning Tool   Fig. 72. Detail of Left Hand Turning Tool

Left Hand Turning Tool

The left hand turning tool illustrated in Figs. 71 and 72 is just the opposite of the right hand turning tool shown in Figs. 69 and 70. This tool is designed for machining work from left to right.

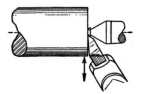

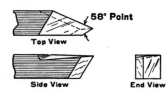

Fig. 73. Application of Right Hand Side Tool   Fig. 74. Detail of Right Hand Side Tool

Right Hand Side Tool

The right hand side tool is intended for facing the ends of shafts and for machining work on the right side of a shoulder. This tool should be fed outward from the center, as indicated by the arrow in Fig. 73. The point of the tool is sharp and is ground to an angle of 58° to prevent interference with the tailstock center. When using this cutter bit care should be taken not to bump the end of the tool against the lathe center, as this will break off the point. See page 29 for correct angle of side clearance and front clearance.

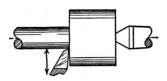

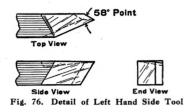

Fig. 75. Application of Left Hand Side Tool   Fig. 76. Detail of Left Hand Side Tool

Left Hand Side Tool

The left hand side tool shown in Figs. 75 and 76 is just the reverse of the right hand side tool shown in Figs. 73 and 74. This tool is used for facing the left side of the work, as shown in Fig. 75.

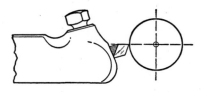

Fig. 77. Application of Screw Thread Cutting Tool   Fig. 78. Detail of Thread Cutting Tool

Thread Cutting Tool

Figs. 77 and 78 show the standard type of cutter bit for cutting United States or American National Form screw threads. The cutter bit is usually ground flat on top, as shown in Fig. 77, and the point of the tool must be ground to an included angle of 60°, as shown in Fig. 78. Careful grinding and setting of this cutter bit will result in perfectly formed screw threads. When using this type of cutter bit to cut screw threads in steel, always keep the work flooded with lard oil in order to obtain a smooth thread. Machine oil may be used if no lard oil is available.

Fig. 79. Application of Brass Turning Tool

Fig. 80. Detail of Brass Turning Tool

Brass Turning Tool

The brass turning tool shown above is similar to the round nosed turning tool illustrated in Figs. 67 and 68, except that the top of the tool is ground flat so that there is no side rake or back rake. This is to prevent the tool from digging into the work and chattering.

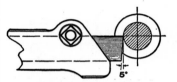

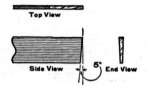

Fig. 81. Application of Cutting-off Tool

Fig. 82. Detail of Cutting-off Tool

Cutting-off Tool

The cutting-off tool should always be set exactly on center, as shown in Fig. 81. This type of tool may be sharpened by grinding the end of the cutter blade to an angle of 5° as shown in Fig. 82. The sides of the blade have sufficient taper to provide side clearance, so do not need to be ground. When cutting off steel always keep the work flooded with oil. No oil is necessary when cutting off cast iron.

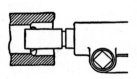

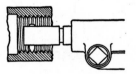

Fig. 83. Application of Boring Tool

Fig. 84. Detail of Boring Tool

Fig. 85. Inside Threading Tool

Boring Tool and Inside Threading Tool

The boring tool is ground exactly the same as the left hand turning tool shown in Figs. 71 and 72 on page 31, except the front clearance of boring tool must be ground at a slightly greater angle so that the heel of the tool will not rub in the hole of the work. The inside threading tool is ground the same as the screw thread cutting tool shown in Figs. 77 and 78 on page 32, except that the front clearance must be increased for the same reason as for the boring tool.

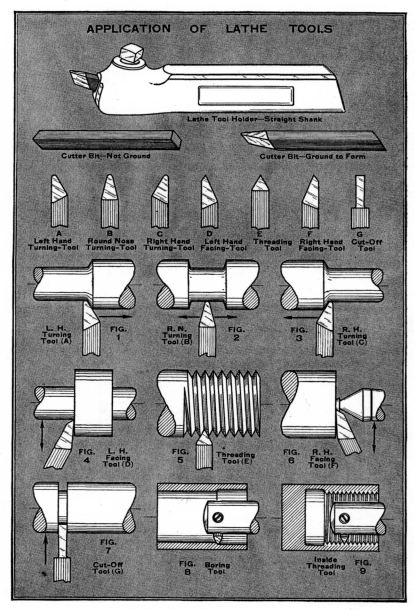

APPLICATION OF LATHE TOOLS

Lathe Tool Holder—Straight Shank

Cutter Bit—Not Ground

Cutter Bit—Ground to Form

A Left Hand Turning-Tool　B Round Nose Turning-Tool　C Right Hand Turning-Tool　D Left Hand Facing-Tool　E Threading Tool　F Right Hand Facing-Tool　G Cut-Off Tool

L. H. Turning Tool (A)　FIG. 1

R. N. Turning Tool (B)　FIG. 2

FIG. 3　R. H. Turning Tool (C)

FIG. 4　L. H. Facing Tool (D)

FIG. 5　Threading Tool (E)

FIG. 6　R. H. Facing Tool (F)

FIG. 7　Cut-Off Tool (G)

FIG. 8　Boring Tool

Inside Threading Tool　FIG. 9

Fig. 86.　Seven of the Most Popular Shapes of Lathe Tool Cutter Bits and Their Application

Stellite Cutter Bits

Stellite cutter bits will stand higher cutting speeds than high speed steel cutter bits. Stellite is also used for machining hard steel, cast iron, bronze, etc.

Fig. 87.   Stellite Cutter Bit

Stellite is a non-magnetic alloy which is harder than ordinary high speed steel. It will stand very high cutting speeds and the tool will not lose its temper even though heated red hot from the friction generated by taking the cut.

Stellite is more brittle than high speed steel, and for this reason should have just enough clearance to permit the tool to cut freely, as the cutting edge must be well supported to prevent chipping and breaking.

Tungsten Carbide Cutting Tools

Tungsten carbide tipped cutting tools are used for manufacturing operations where maximum cutting speeds are desired, and are highly efficient for machining cast iron, alloyed cast iron, copper, brass, bronze, aluminum, babbitt and abrasive non-metallic materials such as fibre, hard rubber and plastics. Cutting speeds may vary from 110 to 650 surface feet per minute, depending on the depth of cut and the feed.

Tungsten carbide tipped cutter bits must be ground on a special grade of grinding wheel, as they are so hard they cannot be satisfactorily ground on the ordinary grinding wheel. The cutting edge must be well supported to prevent chipping and should have just enough clearance to permit the tool to cut freely.

Fig. 88.   Tungsten Carbide Tipped Cutting Tool Mounted in Open Side Tool Post for Rigid Support.

Tantalum Carbide Cutting Tools

Tantalum carbide is a term applied to a combination of tungsten carbide and tantalum carbide. Tantalum carbide tipped cutting tools are similar to tungsten carbide tools, but are used mostly for machining steel.

Titanium Carbide Cutting Tools

Titanium carbide is a term applied to a combination of tungsten carbide and titanium carbide. Titanium carbide is interchangeable with tantalum carbide in its uses.

Fig. 88A.   Machining a Steel Shaft at High Speed with Tantalum Carbide Tipped Cutting Tool.

Cutting Power of Various Sizes of Lathes

The illustrations below show the cutting power of various sizes of lathes and the depth of cut which can be taken on each lathe when the lathe spindle speed is correct and the cutting tools are properly ground and set. The feed used on each cut is approximately .005 in. per revolution of the spindle; the cutting speed is 60 ft. per minute; the metal being machined is .45 Carbon hot rolled steel.

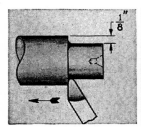

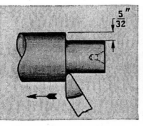

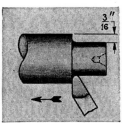

Fig. 90. 9-In. Lathe	Fig. 91. 10-in. Lathes	Fig. 92. 11-In. Lathe
Reducing the diameter ¼ in. in one cut.	Reducing the diameter ⁵⁄₃₂ in. in one cut.	Reducing the diameter ⅜ in. in one cut.

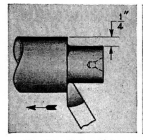

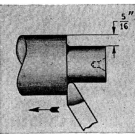

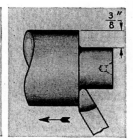

Fig. 93. 13-In. Lathe	Fig. 94. 14½-In. Lathe	Fig. 95. 16-In. Lathe
Reducing the diameter ½ in. in one cut.	Reducing the diameter ⅝ in. in one cut.	Reducing the diameter ¾ in. in one cut.

KEEP YOUR CUTTING TOOLS SHARP

Cutting tools must be ground to a sharp, keen edge in order to do fine, accurate work. Expert mechanics take pride in keeping their tools in first class condition. When you start machining a piece of work, take heavy cuts until within a few thousandths of the finished size, then take light cuts and finish carefully and accurately.

Chapter V

HOW TO TAKE ACCURATE MEASUREMENTS

The ability to take accurate measurements can be acquired only by practice and experience. Careful and accurate measurements are essential to good machine work. All measurements should be made with an accurately graduated steel scale or a micrometer. Never use a cheap steel scale or a wood ruler, as they are likely to be inaccurate and may cause spoiled work.

An experienced mechanic can take measurements with a steel scale and calipers to a surprising degree of accuracy. This is accomplished by developing a sensitive "caliper feel" and by carefully setting the calipers so that they "split the line" graduated on the scale.

Setting an Outside Caliper

A good method for setting an outside caliper to a steel scale is shown in Fig. 96. The scale is held in the left hand and the caliper in the right hand. One leg of the caliper is held against the end of the scale and is supported by the finger of the left hand while the adjustment is made with the thumb and first finger of the right hand.

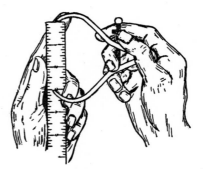

Fig. 96. Setting an Outside Caliper

Measuring with Calipers

The proper application of the outside caliper when measuring the diameter of a cylinder or a shaft is shown in Fig. 97. The caliper is held exactly at right angles to the center line of the work and is pushed gently back and forth across the diameter of the cylinder to be measured. When the caliper is adjusted properly, it should easily slip over the shaft of its own weight. Never force a caliper or it will spring and the measurement will not be accurate.

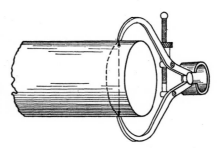

Fig. 97. Measuring with an Outside Caliper

37

Setting Inside Calipers

To set an inside caliper for a definite dimension, place the end of the scale against a flat surface and the end of the caliper at the edge and end of the scale. Hold the scale square with the flat surface. Adjust the other end of the caliper to the required dimension.

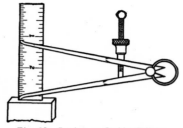

Fig. 98. Setting an Inside Caliper

Measuring Inside Diameters

To measure an inside diameter, place the caliper in the hole as shown on the dotted line and raise the hand slowly. Adjust the caliper until it will slip into the hole with a very slight drag. Be sure to hold the caliper square across the diameter of the hole.

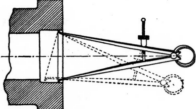

Fig. 99. Measuring with Inside Caliper

Transferring Measurements

In transferring measurement from an outside caliper to an inside caliper, the point of one leg of the inside caliper rests on a similar point of the outside caliper, as shown in Fig. 100. Using this contact point as a pivot, move the inside caliper along the dotted line shown in illustration, and adjust with the thumb screw until you feel your measurement is just right.

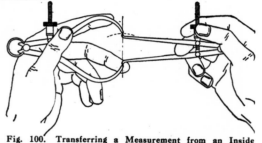

Fig. 100. Transferring a Measurement from an Inside Caliper to an Outside Caliper

Hermaphrodite Caliper

The hermaphrodite caliper shown in Fig. 101 is set from the end of the scale exactly the same as the outside caliper.

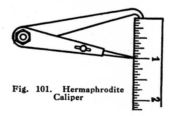

Fig. 101. Hermaphrodite Caliper

Caliper Feel

The accuracy of all contact measurements is dependent upon the sense of touch or feel. The caliper should be delicately and lightly held in the finger tips, not gripped tightly. If the caliper is gripped tightly, the sense of touch is very much impaired.

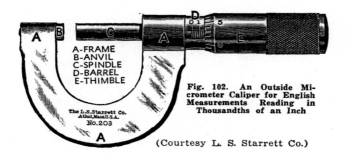

A-FRAME
B-ANVIL
C-SPINDLE
D-BARREL
E-THIMBLE

The L.S.Starrett Co.
Athol,Mass.U.S.A.
No.203

Fig. 102. An Outside Mi-
crometer Caliper for English
Measurements Reading in
Thousandths of an Inch

(Courtesy L. S. Starrett Co.)

How to Read a Micrometer (English Measurement)

Each graduation on the micrometer barrel "D" represents one turn of the spindle or .025 in. Every fourth graduation is numbered and the figures represent tenths of an inch since 4x.025 in. = .100 in. or $\frac{1}{10}$ of an inch.

The thimble "E" has twenty-five graduations, each of which represents one-thousandth of an inch. Every fifth graduation is numbered, from five to 25.

The micrometer reading is the sum of the readings of the graduations on the barrel and the thimble. For example, there are seven graduations visible on the barrel in the illustration above. Since each graduation represents .025 in., the reading on the barrel is 7x.025 in. or .175 in. To this must be added the reading on the thimble which is .003 in. The correct reading is the sum of these two figures or .175 in. + .003 in. = .178 in. Therefore this micrometer is set for a diameter of .178 in.

Metric System Micrometer

Micrometers for measuring in the Metric system are graduated to read in hundredths of a millimeter as shown at right in Fig. 103. For each complete revolution the spindle travels ½ mm or .50 mm, and two complete revolutions are required for 1.00 mm. Each of the upper set of graduations on the barrel represent 1 mm. (two revolutions of the spindle) and every fifth graduation is numbered 0, 5, 10, 15, etc. The lower set of graduations subdivides each millimeter division into two parts.

The beveled edge of the thimble is divided into 50 graduations, each of which represents .01 mm.

The micrometer reading is the sum of the readings on the barrel and the thimble. For example, in Fig. 103 there are three millimeter graduations visible on the barrel, also a ½ mm graduation. The reading on the thimble is 36 mm. Therefore, the reading is 3.00 mm + .50 mm + .36 mm = 3.86 mm.

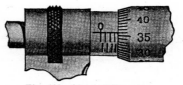

Fig. 103. Metric Micrometer
(Courtesy Brown & Sharpe
Mfg. Co.)

Fig. 104. Testing Headstock Spindle with Test Bar and
Test Indicator

The Accuracy of a Screw Cutting Lathe

In manufacturing the back-geared screw cutting lathe, accuracy is given the most careful attention. A few of the accuracy tests are shown below. The illustration above shows the method of testing the headstock spindle of a lathe to see that the taper of the spindle runs true and that the axis of the spindle is parallel to the ways of the lathe.

The test bar is made of steel and may be from 10 in. to 12 in. long, depending on the size of the lathe. It is machined between centers and ground on the taper shank and also on the two larger diameters where the indicator readings are taken, as shown below. A dial test indicator used with this bar, as shown above, will disclose an error of one ten-thousandth of an inch.

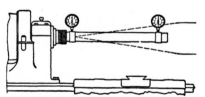

Fig. 105. Testing Alignment of Tailstock
Spindle and Headstock Spindle

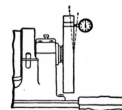

Fig. 106. Testing Amount of Concavity of
Face Plate with Dial Indicator

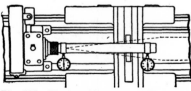

Fig. 107. Testing Alignment of Headstock
Spindle with Ways of Lathe Bed

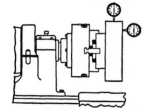

Fig. 108. Testing Accuracy of Chuck Jaws
on Diameter and Face

Fig. 109. Testing Laboratory and Research Department

Testing and Research Laboratory

In a well equipped research laboratory at the South Bend Lathe factory, new ideas, new materials, and new methods are tested. Here measuring instruments and tools are constantly checked to maintain uniform accuracy in South Bend Lathes.

Fig. 110. An Optical Comparator

The equipment of this laboratory includes precision gauge blocks accurate to five-millionths of an inch, an optical comparator for testing the form and lead of screws threads, a profilometer for checking the smoothness of surface finishes, hardness testing equipment to make sure that heat-treated steel surfaces have just the right degree of hardness, precision lead screw testing equipment accurate to .00005" in 30", a dynamic balancing machine, and many other precision instruments, gauges, and tools.

Lathes built today are vastly superior to those of a quarter-century ago. Research in metallurgy has produced steel and iron having greater strength and durability. Better measuring equipment and methods make possible greater precision in the finishing and fitting of machine parts.

Fig. 110-A. Testing Lead Screw for Accuracy

The development of the superfinishing process has resulted in more perfect bearing surfaces.

Since November 1, 1906, the South Bend Lathe Works has been developing methods and equipment for manufacturing precision lathes. Years of careful research have resulted in a continual improvement in South Bend Lathes that has earned them an enviable position of leadership. Today, they are better in every way.

Fig. 111. Using an Outside Micrometer Caliper Measuring the Diameter of
Work in the Lathe

Fig. 112.   Using an Inside Micrometer Caliper Measuring the
Diameter of a Machined Hole

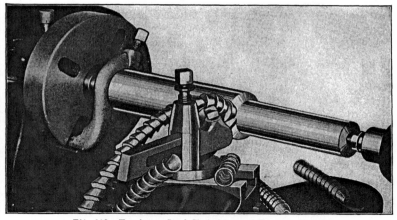

Fig. 113. Turning a Steel Shaft Mounted Between Centers

Chapter VI

PLAIN TURNING

The illustration above shows the lathe in operation machining a shaft between the lathe centers. Whenever possible, work should be mounted in this way for machining as heavier cuts may be taken because the work is supported on both ends.

Locating Center Holes

There are several good methods for accurately locating the center holes which must be drilled in each end of the work before it can be mounted on the lathe centers for machining.

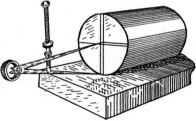

Fig. 114. Locating Centers with Dividers

Divider Method

Chalk the ends of the shaft, set the dividers to approximately one-half the diameter of the shaft, and scribe four lines across each end, as shown in Fig. 114.

Combination Square Method

Hold the center head of a combination square firmly against the shaft, as shown in Fig. 115, and scribe two lines close to the blade across each end of the shaft.

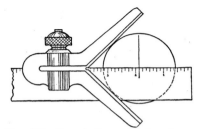

Fig. 115. Use of Center Head to Locate Centers

Hermaphrodite Caliper Method

Chalk each end of the work, set the hermaphrodite caliper a little over half the diameter, and scribe four lines as shown in Fig. 116.

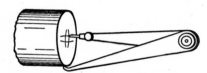

Fig. 116. Centering with Hermaphrodite Calipers

Centering Irregular Shapes

Work that is irregular in shape may be centered with a surface gauge and V-block, as shown in Fig. 117.

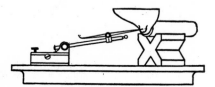

Fig. 117. Centering an Irregular Shape

Bell Center Punch

The bell centering cup is placed over the end of the work and the center punch or plunger is struck a sharp blow with the hammer, automatically locating the center.

Punching the Center

Place the center punch vertically at the center point and tap with a hammer, making a mark sufficiently deep so that the work will revolve on the center points when placed in the lathe.

Test on Centers

After a piece has been center punched it should be tested on centers, as shown in Fig. 120, to make sure that the centers are accurately located. Spin the work with the left hand and mark the high spots on each end of the cylinder with a piece of chalk in the right hand.

Fig. 118. Punching the Center

Fig. 119. Bell Center Punch

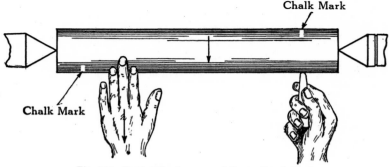

Chalk Mark

Chalk Mark

Fig. 120. Testing the Accuracy of Center Punch Marks

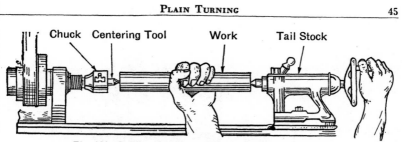

Fig. 121. Drilling the Center Hole in the End of a Shaft

Changing Location of Center Holes

If the centers have not been accurately located, the position of the center punch mark can be changed by placing the center punch at an angle, as shown in Fig. 122, and driving the center over. The shaft should be securely clamped in a vise while this is done.

Drilling the Center Holes

After the centers have been accurately located, center holes should be drilled and countersunk in each end of the shaft. This may be done in the lathe, as shown in Fig. 121, or in a drill press. A combination center drill and countersink, shown in Fig. 124, or a small twist drill followed by a 60° countersink, shown in Fig. 123, may be used.

Fig. 122. Changing the Location of a Center Punch Mark

Center Drill and Countersink

The combination center drill and countersink is usually used for drilling center holes. Several standard sizes suitable for various sizes of work are available, as listed in the tabulation below.

Some care should be exercised in drilling the center holes. The spindle speed should be about 150 r. p. m. and the drill should not be crowded. If the drill is crowded and the point is broken off in the work it may be necessary to heat the end of the shaft to a cherry red and allow it to cool slowly so that the drill point will be annealed and can be drilled out.

Fig. 123. A 60° Countersink

Fig. 124. Combination Center Drill and Countersink

SIZE OF CENTER HOLE FOR $\frac{3}{16}$ IN. TO 4 IN. DIAMETER SHAFTS

Diameter of Work W	Large Diameter of Countersunk Hole C	Diameter of Drill D	Diameter of Body F
$\frac{3}{16}$ in. to $\frac{5}{16}$ in.	$\frac{1}{8}$ in.	$\frac{1}{16}$ in.	$\frac{13}{64}$ in.
$\frac{3}{8}$ in. to 1 in.	$\frac{3}{16}$ in.	$\frac{3}{32}$ in.	$\frac{3}{16}$ in.
$1\frac{1}{4}$ in. to 2 in.	$\frac{1}{4}$ in.	$\frac{1}{8}$ in.	$\frac{3}{16}$ in.
$2\frac{1}{4}$ in. to 4 in.	$\frac{5}{16}$ in.	$\frac{5}{32}$ in.	$\frac{7}{16}$ in.

Drilling Center Holes with a Lathe Chuck

Small diameter rods that can be passed through the headstock spindle and short shafts are easily centered with the aid of a universal chuck, as shown in Fig. 125. When this method is used, the end of the shaft should be faced smooth before drilling the center hole.

The unsupported end of the shaft should not extend more than 10 in. beyond the chuck jaws. Shafts that are too large to pass through the headstock and too long to be held firmly by the chuck jaws alone can be supported on the outer end in a center rest. (See page 92.)

Correct Center Hole

To be correct, the center hole must be the size required for the diameter of the shaft, as listed in the tabulation on page 45, and the countersink must fit the center point perfectly, as shown in Fig. 126. There must also be sufficient clearance at the bottom of the countersink.

When drilling center holes, allow for the thickness of the metal that will be faced off of the end; otherwise, the center holes will be too small to support the shaft after the ends are faced.

Poorly Drilled Center Holes

One of the most common causes of unsatisfactory lathe work is poorly drilled center holes. Fig. 127 shows a shallow center hole with incorrect angle and no clearance for the tip of the center point. Fig. 128 shows a center hole that has been drilled too deep. Accuracy cannot be expected when center holes are poorly made, and the lathe centers may be damaged.

Fig. 125. Drilling a Center Hole with Stock Held in Lathe Chuck and Drill in Tailstock Spindle Chuck

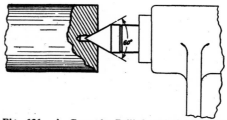

Fig. 126. A Correctly Drilled and Countersunk Hole Fits the Lathe Center Perfectly

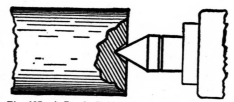

Fig. 127. A Poorly Drilled Center Hole, Too Shallow and Incorrect Angle

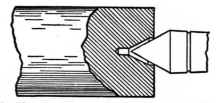

Fig. 128. An Incorrect Center Hole, Drilled Too Deep to Fit Center

Fig. 129. Standard Lathe Dog

Fig. 130. Safety Lathe Dog

Lathe Dogs for Driving Work on Centers

The common lathe dog shown in Fig. 129 is the most popular type. Fig. 130 shows a safety lathe dog which has a headless set screw and is not likely to catch in the operator's sleeve. Fig. 132 shows a clamp lathe dog, used principally for rectangular work in the lathe. When attaching the lathe dog to the work, make sure that the set screw is securely tightened.

Mounting Centers in the Lathe Spindles

Before mounting the lathe centers in the headstock or tailstock spindle, thoroughly clean the centers, the tapered holes and the spindle sleeve "A," Fig. 131. A very small chip or a little dirt will cause the center to run out of true. Use cloth and stick to clean taper hole. Do not insert finger in revolving spindle.

The tail spindle center is hardened and tempered and is marked with a groove to distinguish it from the head spindle center.

Removing the Lathe Centers

With a piece of rag in your right hand, hold the sharp point of the headstock center, and with the left hand give the center a sharp tap with a rod through the spindle hole. Fig. 133 shows a steel rod with a small bushing attached for removing the headstock spindle center and taper sleeve.

To remove the tailstock center, turn the tailstock hand wheel to the left until the end of the tailstock screw bumps the end of the center. This will loosen the center and it may be removed from the spindle.

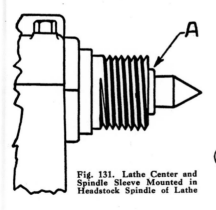

Fig. 131. Lathe Center and Spindle Sleeve Mounted in Headstock Spindle of Lathe

Fig. 132. Clamp Lathe Dog

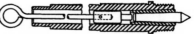

Fig. 133. Knock Out Bar for Removing Headstock Spindle Center

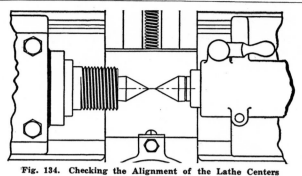

Fig. 134. Checking the Alignment of the Lathe Centers

Check Alignment of Centers

Before mounting work between the lathe centers they should be checked for alignment, as shown in Fig. 134. If the tailstock center does not line up, loosen the tailstock clamp bolt and set over the tailstock top in the proper direction by adjusting the tailstock set over screws. (See page 51.)

Mounting the Work Between Centers

Place a drop of oil in the center hole for the tailstock center point before mounting the work between centers. The tail of the lathe dog should fit freely into the slot of the face plate so that the work rests firmly on both the headstock center and the tailstock center, as shown in Fig. 136. Make sure that the lathe dog does not bind in the slot of the face plate, as shown in Fig. 135.

The tailstock center should not be tight against the work, but should not be too loose. The work must turn freely, for if the tail center is too tight it will stick and may be ruined.

Expansion of Work

When work is machined in the lathe it may become hot and expand. The expansion of work mounted between centers will cause it to bind, making it necessary to stop the lathe and readjust the tailstock center. When machining a long shaft, several readjustments of the tailstock center may be required.

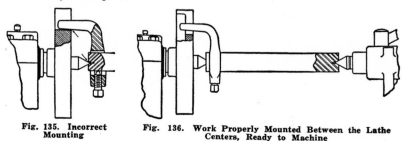

Fig. 135. Incorrect   Fig. 136. Work Properly Mounted Between the Lathe
Mounting   Centers, Ready to Machine

Facing the Ends

Before turning the diameter of a shaft, the ends should be faced square. Grind the cutter bit as shown in Fig. 74, page 32, and set the cutting edge exactly on center, as shown in Fig. 137. Be careful not to break the point of the tool against the tailstock center. Feed the tool out to face the end, as shown in Fig. 138.

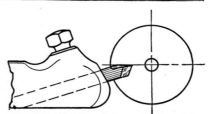

Fig. 137. Position of Cutter Bit for Facing the End of a Shaft

Position of Tool for Turning

Grind the cutter bit for turning as shown in Fig. 64, page 30. The cutting edge of the cutter bit and the end of the tool holder should not extend over the edge of the compound rest any farther than necessary. (See "A" and "B," Fig. 139.)

The tool should be set as shown in Fig. 140 so that if the tool slips in the tool post it will not dig into the work, but instead it will move in the direction of the arrow away from the work.

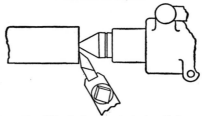

Fig. 138. Facing the End of a Shaft

Direction of Power Feed

The feed of the tool should be toward the headstock, if possible, so that the pressure of the cut is on the head spindle center which revolves with the work.

Rate of Power Feed

The rate of the power feed depends on the size of the lathe, the nature of the work, and the amount of stock to be removed.

On a small lathe a feed of .008 in. per revolution of the spindle may be used, but on larger sizes of lathes feeds as coarse as .020 in. are often used for rough turning. Care must be taken when turning long slender shafts as a heavy cut with a coarse feed may bend the shaft and ruin the work.

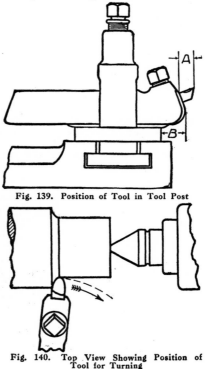

Fig. 139. Position of Tool in Tool Post

Fig. 140. Top View Showing Position of Tool for Turning

Cutting Speeds for Turning

The most efficient cutting speed for turning varies with the kind of metal being machined, the depth of the cut, the feed and the type of cutter bit used. If too slow a cutting speed is used, much time may be lost, and if too high a speed is used the tool will dull quickly. The following cutting speeds are recommended for high speed steel cutter bits:

CUTTING SPEEDS IN SURFACE FEET PER MINUTE

Kind of Metal	Roughing Cuts .010 in. to .020 in. Feed	Finishing Cuts .002 in. to .010 in. Feed	Cutting Screw Threads
Cast Iron	60 f. p. m.	80 f. p. m.	25 f. p. m.
Machine Steel	90 f. p. m.	100 f. p. m.	35 f. p. m.
Tool Steel, Annealed ...	50 f. p. m.	75 f. p. m.	20 f. p. m.
Brass	150 f. p. m.	200 f. p. m.	50 f. p. m.
Aluminum	200 f. p. m.	300 f. p. m.	50 f. p. m.
Bronze	90 f. p. m.	100 f. p. m.	25 f. p. m.

If a cutting lubricant is used, the above speeds may be increased 25% to 50%. When using tungsten-carbide tipped cutting tools, the cutting speeds may be increased from 100% to 800%.

To find the number of revolutions per minute required for a given cutting speed, in feet per minute, multiply the given cutting speed by 12 and divide the product by the circumference (in inches) of turned part.

Example: Find the number of revolutions per minute for 1 in. shaft for a cutting speed of 90 ft. per minute.

$$\frac{90 \times 12}{3.1416 \times 1} = 343.77 \text{ r. p. m.}$$

Spindle speeds for various diameters and metals are listed in the tabulation below to eliminate the necessity of making calculations. Refer to page 23 for spindle speeds of various sizes of lathes.

SPINDLE SPEEDS IN R.P.M. FOR TURNING AND BORING
Calculated for Average Cuts with High Speed Steel Cutter Bits

Diameter in Inches	Alloy Steels 50 f. p. m.	Cast Iron 75 f. p. m.	Machine Steel 100 f. p. m.	Hard Brass 150 f. p. m.	Soft Brass 200 f. p. m.	Aluminum 300 f. p. m.
1	191	287	382	573	764	1146
2	95	143	191	287	382	573
3	64	95	127	191	254	381
4	48	72	95	143	190	285
5	38	57	76	115	152	228
6	32	48	64	95	128	192
7	27	41	55	82	110	165
8	24	36	48	72	96	144
9	21	32	42	64	84	126
10	19	29	38	57	76	114
11	17	26	35	52	70	105
12	16	24	32	48	64	96
13	15	22	29	44	58	87
14	14	20	27	41	54	81
15	13	19	25	38	50	75
16	12	18	24	36	48	72

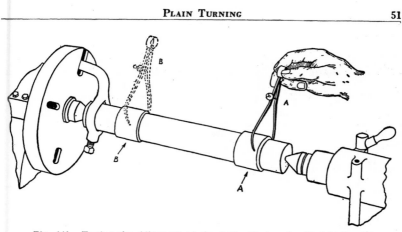

Fig. 141. Testing the Alignment of the Lathe Centers for Straight Turning

Testing Alignment of Centers

After taking the first roughing cut across a shaft, check the diameter at each end of the cut with calipers or micrometers to make sure the lathe is turning straight. Sometimes when the position of the tailstock is changed for a different length of work, it will require adjustment. This is especially true on old lathes which may have worn spots and burrs on the bed.

Fig. 141 shows a good method of testing for the alignment of the lathe centers. Two collars, A and B, turned on a shaft about 1½ in. in diameter and 10 in. long, are machined with a fine finishing cut without changing the adjustment of the cutting tool. Collar A is measured, and without making any adjustment on the caliper, collar B is tested to see how it compares with collar A. If collar A is not the same diameter as collar B, then the adjustment of centers is not correct, and the tailstock top should be adjusted in the direction required.

Adjustment of Tailstock Top

The tailstock top is adjusted by releasing one of the adjusting screws of the tailstock top and tightening the opposite screw a similar distance. Then take another test cut on the collars, measure and continue this operation until the desired degree of accuracy is obtained.

There is a mark on the end of the tailstock where the bottom and top join to show the relative position of the tailstock top and bottom. For fine, accurate work, this mark should not be depended upon, but the alignment test should be made as described above to be sure that the centers are in line.

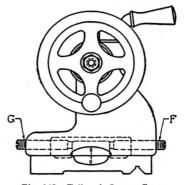

Fig. 142. Tailstock Set on Center

Machining to a Shoulder

A good method for locating a shoulder on a shaft is shown in Fig. 143. After chalking the shaft, set the hermaphrodite calipers to the required dimension and scribe a line around the revolving shaft with the sharp point on the caliper.

Fig. 144 shows the use of a round nosed turning tool for finishing a shoulder having a fillet corner. (See page 31, Fig. 67.)

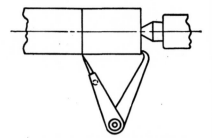

Fig. 143. Measuring with Hermaphrodite Caliper to Locate a Shoulder

Locating Shoulders with a Parting Tool

In production work where a quantity of pieces are required, shoulders are usually located with a parting tool, as shown in Fig. 146, before the diameter is machined.

When a square corner is required, as for a bearing, it is customary to neck or undercut the shoulder slightly, as shown in Fig. 145.

A firm joint caliper is convenient for measuring when facing the ends of a shaft to length or for measuring between two shoulders, as shown in Fig. 147.

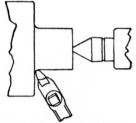

Fig. 144. Finishing a Shoulder with a Fillet

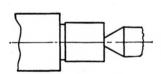

Fig. 145. Detail of a Shoulder with a Recess

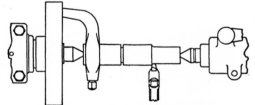

Fig. 146. Location of a Shoulder Marked with a Parting Tool Before Turning the Diameter

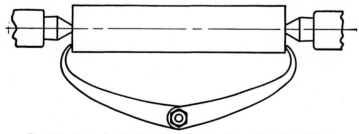

Fig. 147. Measuring the Length of a Shaft with a Firm Joint Caliper

Fig. 148. Machining a Shaft in an Independent Chuck

Chapter VII

CHUCK WORK

Work that cannot readily be mounted between the lathe centers is usually held in a chuck, as shown above, for machining. Several types of chucks are used, but the most popular are the 4-jaw Independent chuck and the 3-jaw Universal chuck shown below.

A 4-jaw Independent chuck has four reversible jaws, each of which may be independently adjusted. This type of chuck is recommended if the lathe is to have but one chuck, as it will hold square, round and irregular shapes in either a concentric or an eccentric position.

The 3-jaw Universal chuck is used for chucking round and hexagonal work quickly, as the three jaws move simultaneously and automatically center the work. Two sets of jaws are required, one set for external chucking and the other for internal chucking.

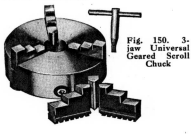

Fig. 149. 4-jaw Independent Chuck

Fig. 150. 3-jaw Universal Geared Scroll Chuck

53

Mounting Chuck on Spindle

Before mounting a chuck or a face plate on the lathe spindle, thoroughly clean and oil the threads of the spindle and the chuck back. Also clean the shoulder of the spindle where the chuck back fits against it. A very small chip or burr at this point will prevent the chuck from running true.

Hold the chuck against the spindle nose with the right hand and arm and turn the lathe spindle cone with the left hand to screw the chuck onto the spindle just tight enough to hold it securely.

Do not run the lathe with power while screwing the chuck onto the spindle, and do not spin the chuck up to the shoulder or it may be very difficult to remove.

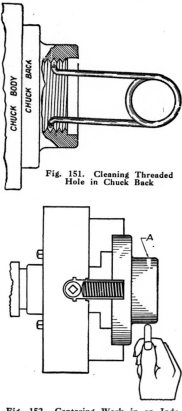

Fig. 151. Cleaning Threaded Hole in Chuck Back

The Independent Chuck

The independent chuck is used more than any other type of chuck because it will hold practically any shape and can be adjusted to any degree of accuracy required.

Concentric rings scribed on the face of the chuck permit centering round work approximately, as it is placed in the chuck. To center more accurately, the lathe is started and a piece of chalk is held lightly against the revolving work, as shown in Fig. 152. The lathe is then stopped and the jaw opposite the chalk mark is loosened slightly. The opposite jaw is then tightened. The test is repeated until the work is centered with the necessary accuracy. All four jaws must be securely tightened before starting to machine the work.

Fig. 152. Centering Work in an Independent Chuck

Use of Center Indicator

The center indicator is used for accurately centering work that has been laid out and center punched for drilling and boring. The short end of the center indicator is placed in the center punch mark and the tailstock center point is brought up close to the opposite end, as shown in Fig. 153. For accurate work, the long end of the indicator should remain stationary when the lathe spindle is revolved.

Fig. 153. Centering Work with a Center Indicator

Centering Work With Dial Test Indicator

A sensitive dial indicator may be used for accurately centering work having a smooth surface. The dial of the indicator is graduated to read in thousandths of an inch so that practically any required degree of accuracy may be obtained.

Fig. 154. Centering Work with a Dial Test Indicator

The indicator is placed in contact with the part to be centered, as shown in Fig. 154, and the hand on the indicator dial is watched as the lathe spindle is revolved slowly by hand. The chuck jaws are adjusted as described on page 54 until the required degree of accuracy is obtained.

The part to be centered should also be tested on the face for wobble as shown in Fig. 155.

Removing Chuck from Lathe Spindle

An easy way to remove a chuck or a face plate from the lathe spindle is shown in Fig. 156. The back gears are engaged and the belt is placed on the large step of the cone pulley. A block of wood is placed between the chuck jaw and the back V-way of the lathe bed and the belt is pulled by hand as shown.

Fig. 155. Testing Face of Work with Dial Indicator for Wobble

Practical Sizes of Chucks

Lathe chucks should be carefully selected for the size of the lathe and the work for which they are to be used. If the chuck is too small, the capacity of the lathe is restricted, but if it is too large the jaws may strike the lathe bed and the chuck will be awkward to use and difficult to handle.

The most practical sizes of chucks for use with various sizes of lathes are listed in the tabulation below.

Fig. 156. Removing a Chuck from Lathe Spindle

Practical Sizes of Chucks for Lathes

Size of Lathe	4-Jaw Independent Lathe Chuck	3-Jaw Universal Geared Scroll Chuck
9-in. lathe.............	6 in.	5 in.
10-in. lathe.............	6 in.	5 in.
13-in. lathe.............	7½ in.	6 in.
14½-in. lathe.............	9 in.	7½ in.
16-in. lathe.............	10 in.	7½ in.

The Universal Chuck

Round and hexagonal work may be chucked quickly in the universal chuck as all three jaws move simultaneously and automatically center the work within a few thousands of an inch. This type of chuck will usually center work within .003 in. when new, but when the scroll becomes worn this degree of accuracy cannot be expected.

Since there is no way to adjust the jaws independently on this type of chuck, it is not used where extreme accuracy is required. The 4-jaw independent chuck should always be used when work must be centered to run dead true. However, if no independent chuck is available shims may be placed between the chuck jaws and the work to compensate for the inaccuracy of universal chuck.

Fig. 157.   Round Work Held in a Universal Chuck

Fig. 158.   Hollow Headstock Spindle Chuck

Headstock Spindle Chuck

The headstock spindle chuck, shown in Figs. 158 and 159, is similar to a drill chuck except that it is hollow and is threaded so that it may be screwed on to the spindle nose of the lathe.

This type of chuck is suitable for holding bars, rods and tubes that are passed through the headstock spindle of the lathe, also other small diameter work. It is more accurate than the average universal chuck and will usually center work within .002 in.

The headstock spindle chuck is inexpensive and for some classes of work it may be used instead of the more expensive draw-in collet chuck.

Fig. 159.   Machining a Shaft in a Hollow Headstock Chuck

Drill Chuck

Drill chucks are used in both the headstock spindle and the tailstock spindle of the lathe for holding drills, reamers, taps, etc. There are several types of drill chucks on the market and some do not have sufficient accuracy and holding power for satisfactory use on the lathe. A good drill chuck will hold drills concentric within .002 in. or .003 in. and should have a wrench or a pinion key for tightening.

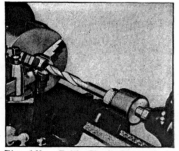

Fig. 160.   Drill Chuck Mounted in Tailstock Spindle

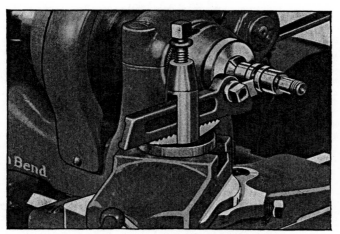

Fig. 161. Machining a Part Held in a Draw-in Collet Chuck

Draw-in Collet Chuck

The draw-in collet chuck is the most accurate of all types of chucks and is used for precision work, such as making small tools and manufacturing small parts for watches, typewriters, radios, etc. The collets are made for round, square and other shapes, as shown in Figs. 162, 164 and 165. The work held in the collet should not be more than .001 in. smaller or .001 in. larger than the collet size. If the diameter of the work varies more than this, it will impair the accuracy and efficiency of the collet. A separate collet should be used for each diameter of the work.

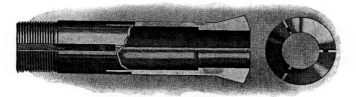

Fig. 162. Side View and End View of a Spring Collet for Round Work

Fig. 163. Handwheel Draw-in Collet Chuck

Fig. 164.
Square Collet

Fig. 165.
Hexagon Collet

Fig. 166.   Cross-Section of Headstock Showing Construction of Draw-in
Collet Chuck Attachment

Construction of Draw-in Collet Chuck

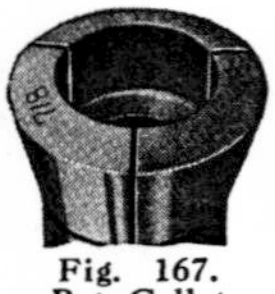

The construction of the draw-in collet chuck is shown above in Fig. 166. The hollow draw-bar with handwheel attached extends through the headstock spindle of the lathe and is threaded on the right end to receive the spring collet. Turning the handwheel to the right draws the collet into the tapered closing sleeve and tightens the collet on the work. Turning the handwheel to the left releases the collet.

Fig. 167.
Pot Collet

Step Chuck and Closer

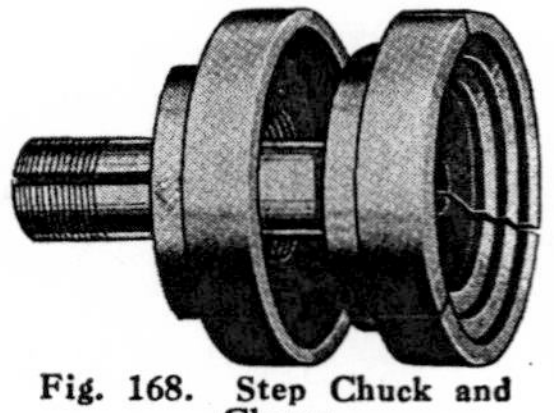

The spring collet may be replaced with a step chuck and closer, as shown in Fig. 168, for holding discs such as gear blanks, etc. A pot collet, shown in Fig. 167, may be used in place of the step chuck for small diameters.

Fig. 168.   Step Chuck and Closer

Handlever Draw-in Collet Chuck

The handlever draw-in collet chuck, shown in Fig. 169, is similar to the handwheel type draw-in collet chuck except that the collet is opened and closed by moving the hand lever to the right or left. This permits gripping or releasing the work without stopping the lathe spindle if desired.

Fig. 169.   Handlever Type Draw-in Collet
Chuck Attachment

Fig. 170. A Set of Collets for Round Work $\frac{1}{16}$ In. to $\frac{3}{4}$ In. by 16ths

Chapter VIII

TAPER TURNING AND BORING

There are three methods of turning and boring tapers in the lathe: by setting over the tailstock; by using the compound rest; and by using the taper attachment of the lathe. The method used depends on the length of the taper, the angle of taper and the number of pieces to be machined.

Cutter Bit Must Be on Center

The cutting edge of the tool must be set exactly on center, as shown in Fig. 173, to turn or bore an accurate taper. That is, the cutting edge of the lathe tool must be exactly the same height as the point of the tailstock center. The position of the tool applies to all methods of turning and boring tapers.

Taper Turning with Compound Rest

The compound rest of the lathe is usually used for turning the boring short tapers and bevels, especially for bevel gear blanks and for die and pattern work, etc. The compound rest swivel is set at the required angle and the taper is machined by turning the compound rest feed screw by hand. See Figs. 171 and 172.

Fig. 171. Machining a Conical Punch and Die with Compound Rest

Fig. 172. Compound Rest Set at an Angle for Taper Turning and Boring

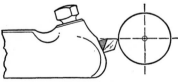

Fig. 173. Lathe Tool Cutter Bit Set on Center for Taper Turning

Truing a 60° Center

A good example of taper turning with the compound rest is machining a 60° center point, as shown in Fig. 174.

The tailstock center is hardened and must be annealed before it can be machined, although it can be trued with a grinding attachment in place of the lathe tool.

Testing the Angle

All angular or bevel turning should be tested with a gauge of some kind, as it is difficult to read the graduations with sufficient accuracy to set the compound rest swivel for an exact taper. Fig. 175 shows the use of a center gauge for testing the 60° angle of a lathe center point.

Taper Turning with Tailstock Set Over

Work that can be machined between centers can be turned taper by setting over the tailstock top, as shown in Figs. 177, 178 and 179. This method cannot be used for boring tapers.

The amount the tailstock top must be set over depends on the amount of the taper per foot and the over-all length of the work. With the same amount of setover, pieces of different lengths will be machined with different tapers, as shown in Fig. 177. Notice that the tailstock center is set over one-half the total amount of the taper for the entire length of the work.

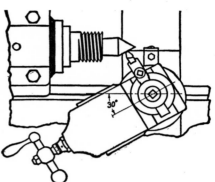

Fig. 174. Machining a 60° Lathe Center Point

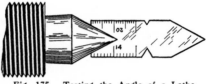

Fig. 175.   Testing the Angle of a Lathe Center Point

Fig. 176. Machining a Taper with the Tailstock Top Set Over

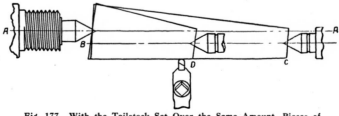

Fig. 177. With the Tailstock Set Over the Same Amount, Pieces of Different Lengths are Machined with Different Tapers

How to Calculate Amount of Setover for Tailstock

Tapers are usually specified in "inches per foot." For example, some Brown & Sharpe tapers are ½ in. per foot. To machine a Brown & Sharpe taper of ½ in. per foot on a shaft exactly one foot long, the center should be set over ¼ in. or .250 in. If the piece is only 10 in. long, then the amount of setover would be $\frac{10}{12}$ of .250 in. or .208 in. The following rules may be used for calculating the amount of setover:

TAPER IN INCHES PER FOOT GIVEN—Divide the total length of the stock in inches by twelve and multiply this quotient by one-half the amount of taper per foot specified. The result is the amount of setover in inches.

DIAMETERS AT ENDS OF TAPER GIVEN—Divide the total length of the stock by the length of the portion to be tapered and multiply this quotient by one-half the difference in diameters; the result is the amount of the setover.

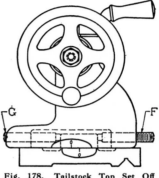

Adjusting the Tailstock Center

To set over the tailstock center for taper turning, loosen clamping nut of tailstock and back off set screw "G," Fig. 178, the distance required; then screw in set screw "F" a like distance until it is tight, and clamp the tailstock to the lathe bed.

Fig. 178. Tailstock Top Set Off Center for Taper Turning

Measuring the Setover

To measure the setover of the tailstock center, place a scale having graduations on both edges between the two centers, as shown in Fig. 179. This will give an approximate measurement.

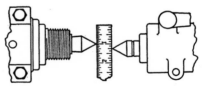

Fig. 179. Measuring the Setover

Fitting Tapers to Gauges

The best way to machine an accurate taper is to fit the taper to a standard gauge. To test the taper, make a chalk mark along the entire length of the taper, place the work in the taper it is to fit and turn carefully by hand. Then remove the work and the chalk mark will show where the taper is bearing.

If the taper is a perfect fit, it will show along the entire length of the chalk mark. If the taper is not perfect, make the necessary adjustment, take another light cut and test again. Be sure the taper is correct before turning to the finished diameter.

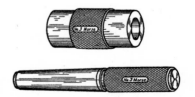

Fig. 180. Morse Standard Taper Plug Gauge and Socket Gauge

Taper Turning with Taper Attachment

The taper attachment is used for turning and boring tapers in the lathe. It eliminates the necessity of setting over the tailstock, and if desired may be set permanently for a standard taper. The taper attachment does not interfere with using the lathe for straight turning.

The taper attachment is especially valuable for boring tapered holes. If the lathe is not equipped with a taper attachment, the compound rest top swivel may be set for the desired taper, but the length of the taper is limited to the comparatively short angular feed of the compound rest top when this method is used.

Graduations on one end of the taper attachment swivel bar indicate the total taper in inches per foot, and on the other end, the included angle of the taper is shown in degrees. See Fig. 182-A.

Plain Taper Attachment

The plain taper attachment (used on all 9″ South Bend Lathes) shown at right consists of a bracket attached to the back of the lathe carriage, a compound slide with clamp for locking slide to lathe bed and a connecting bar to connect the slide block of the taper attachment to the compound rest base of the lathe.

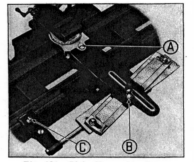

When the plain taper attachment is to be used it is necessary to disconnect the cross feed screw by removing the bolt "A" which locks the cross feed nut to the compound rest base of the lathe. This leaves the compound rest base free to slide so that it may be controlled by the taper attachment. Binding screws "B" and "C" are tightened to engage the taper attachment.

Fig. 181. Plain Taper Attachment

Telescopic Taper Attachment

The telescopic taper attachment (used on 10″ to 16″ swing South Bend Lathes) shown in the illustration below, Fig. 182, is similar to the plain taper attachment described above except that it is equipped with a telescopic cross feed screw. This feature eliminates the necessity of disconnecting the cross feed screw when the taper attachment is to be used.

The cross feed screw may be used to adjust the turning tool for the required diameter and the taper attachment may then be engaged by tightening binding screws "X" and "Y". To change back to straight turning, it is necessary to loosen binding screws "X" and "Y".

Fig. 182. Telescopic Taper Attachment

Setting the Taper Attachment Swivel Bar

Tapers are usually specified in inches per foot or in degrees. When this information is not available the taper in inches per foot should be calculated before setting the taper attachment for taper turning.

To calculate the taper in inches per foot, subtract the diameter in inches at the small end of taper (B, Fig. 183) from the diameter in inches at the large end of the taper (A); divide the result by the length of tapered portion (C) in inches, and multiply by 12. The answer is the taper in inches per foot and indicates the

Fig. 182A. Graduations in Inches per Foot on Swivel Bar

graduation on the "taper per foot" end of the swivel bar which should be set in line with the witness mark in order to machine the desired taper. See Fig. 182A.

Each graduation on the "taper per foot" end of swivel bar represents a total taper of 1/16″ per foot. If the taper per foot has been calculated or specified in decimal fractions instead of common fractions, refer to the decimal equivalent table on page 115 for the nearest fractional part of an inch.

When setting the swivel bar for taper turning, remember that the total taper is indicated by the graduations, either in inches per foot or in degrees. For example, if the swivel bar is set at 5° the taper machined will have a total included angle of 5°; that is 2½° (not 5°) each side of center line.

After setting the taper attachment swivel bar to the required angle, take a trial cut and test the taper with a taper gage or micrometers. Some readjustment of the swivel bar will probably be necessary, as it is difficult to align the graduations on swivel bar perfectly with the witness mark. See page 61 for information on fitting and testing tapers with taper gauges.

Standard tapered holes may be hand reamed after boring to standardize the taper and size the hole.

Turning an outside taper is shown in Fig. 183. Boring an inside taper with the aid of a center rest is shown in Fig. 184. See page 92.

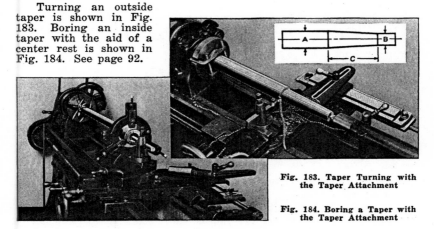

Fig. 183. Taper Turning with the Taper Attachment

Fig. 184. Boring a Taper with the Taper Attachment

Morse Standard Tapers

Morse Standard Tapers are used for lathe and drill press spindles by most of the manufacturers of lathes and drill presses in the United States. South Bend Lathes have both head and tailstock spindles fitted for Morse Standard Tapers. The dimensions of various sizes of Morse Standard Tapers are listed in the tabulation below.

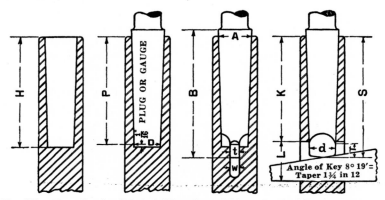

Fig. 185.   Chart Showing Principal Dimensions of Morse Standard Tapers Which are Listed in Tabulation Below

DIMENSIONS OF MORSE STANDARD TAPERS
All Dimensions in Tabulations Below Are in Inches

Number of Taper	Diam. of Plug at Small End	Diam. at End of Socket	Shank		Depth of Hole	Standard Plug Depth	Tongue		Keyway		End of Socket to Keyway	Taper per Foot
			Whole Length	Depth			Thickness	Length	Width	Length		
	D	A	B	S	H	P	t	T	W	L	K	
0	0.252	0.3561	$2\frac{11}{32}$	$2\frac{7}{32}$	$2\frac{1}{32}$	2	$\frac{5}{32}$	$\frac{1}{4}$	0.160	$\frac{9}{16}$	$1\frac{15}{16}$	.6246
1	0.369	0.475	$2\frac{9}{16}$	$2\frac{7}{16}$	$2\frac{3}{16}$	$2\frac{1}{8}$	$\frac{13}{64}$	$\frac{3}{8}$	0.213	$\frac{3}{4}$	$2\frac{1}{16}$	.5986
2	0.572	0.700	$3\frac{1}{8}$	$2\frac{15}{16}$	$2\frac{5}{8}$	$2\frac{9}{16}$	$\frac{1}{4}$	$\frac{7}{16}$	0.260	$\frac{7}{8}$	$2\frac{1}{2}$	.5994
3	0.778	0.938	$3\frac{7}{8}$	$3\frac{11}{16}$	$3\frac{1}{4}$	$3\frac{3}{16}$	$\frac{5}{16}$	$\frac{9}{16}$	0.322	$1\frac{3}{16}$	$3\frac{1}{16}$	.6023
4	1.020	1.231	$4\frac{7}{8}$	$4\frac{5}{8}$	$4\frac{1}{8}$	$4\frac{1}{16}$	$\frac{15}{32}$	$\frac{5}{8}$	0.478	$1\frac{1}{4}$	$3\frac{7}{8}$	.6233
5	1.475	1.748	$6\frac{1}{8}$	$5\frac{7}{8}$	$5\frac{1}{4}$	$5\frac{3}{16}$	$\frac{5}{8}$	$\frac{3}{4}$	0.635	$1\frac{1}{2}$	$4\frac{15}{16}$	.6315
6	2.116	2.494	$8\frac{9}{16}$	$8\frac{1}{4}$	$7\frac{3}{8}$	$7\frac{1}{4}$	$\frac{3}{4}$	$1\frac{1}{8}$	0.760	$1\frac{3}{4}$	7	.6256
7	2.750	3.270	$11\frac{5}{8}$	$11\frac{1}{4}$	$10\frac{1}{8}$	10	$1\frac{1}{8}$	$1\frac{3}{8}$	1.135	$2\frac{5}{8}$	$9\frac{1}{2}$	.6240

The figures in the "Taper per Foot" column have been revised to conform with the standard end diameters and lengths.

Brown & Sharpe and Jarno Tapers

Two other system of tapers are widely used. The Brown & Sharpe Tapers are used for milling machine spindles and the Jarno Tapers for some makes of lathe spindles. Specifications of these tapers can be obtained from technical hand books or from manufacturers using them.

Chapter IX

DRILLING, REAMING AND TAPPING

Many drilling, reaming and tapping jobs can be done more quickly and with greater accuracy in the lathe than by any other method.

Fig. 186 (at right) illustrates the use of the lathe as a drill press. A drill pad placed in the tailstock spindle of the lathe is used to support the work.

The tailstock hand wheel is turned as the hole is drilled through the work. The end of the work may rest on the lathe bed if desired.

Fig. 186. Using the Lathe as a Drill Press

The Location of the hole should be center punched to start the drill. The lathe should be operated at high speed when drilling small diameter holes.

Drill Pad for Tailstock

A drill pad for the tailstock spindle of the lathe is shown in Fig. 187. The drill pad replaces the tailstock center and supports the work for drilling.

Crotch Center

The crotch center shown in Figs. 187 and 188 is similar to the drill pad except that it has a "V" so that round work may be accurately cross drilled. This is very convenient for drilling oil holes in bushings, drilling pin holes in shafts, etc.

Fig. 187. (Right) Crotch Center for Use in Tailstock of Lathe

Fig. 187-A. (Left) Drill Pad for Use in Tailstock of Lathe

Fig. 188. Drilling an Oil Hole in a Bushing with Crotch Center in Tailstock

Drilling Work Held in the Chuck

Most of the drilling in the lathe is done with the work mounted in the lathe chuck (as shown in Fig. 189) or clamped to the face plate of the lathe. When this method is used it is important that the drill be started so that it will run true and the hole will be drilled concentric with the outside diameter of the work.

One method for starting the drill point true is illustrated in Fig. 190. The butt end of a lathe tool holder just touching the side of the drill will prevent the drill from bending and cause it to start approximately true in the center of the work.

Center Drilling

When greater accuracy is required it is best to provide a true starting point for the drill. To do this the work should first be center drilled using a combination center drill and countersink, as shown in Fig. 191. The point of the center drill may be ground off as shown in Fig. 192 to prevent breaking.

Flat Centering Drill

A flat centering tool or drill of forged steel held rigidly in the tool post of the lathe, as shown in Figs. 193 and 194, is often used for production drilling. This type of centering drill provides an accurate starting point for the drill, and since the centering drill is mounted in the tool post, it may be moved back out of the way for the drill which is to follow. This saves time as it eliminates the necessity of changing drills in the tailstock spindle.

Drilling in Steel

When drilling in steel use plenty of lard oil on the point of the drill. If no lard oil is available, any good cutting oil or even machine oil may be used. However, lard oil is preferable, and for some deep hole drilling is the only satisfactory lubricant.

Fig. 189. Drilling Work Held in Chuck

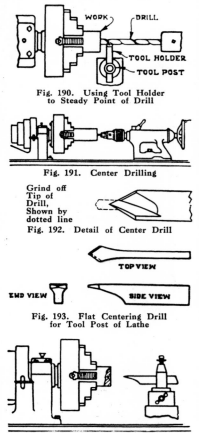

Fig. 190. Using Tool Holder to Steady Point of Drill

Fig. 191. Center Drilling

Grind off Tip of Drill, Shown by dotted line

Fig. 192. Detail of Center Drill

TOP VIEW

END VIEW　　SIDE VIEW

Fig. 193. Flat Centering Drill for Tool Post of Lathe

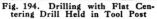

Fig. 194. Drilling with Flat Centering Drill Held in Tool Post

Drilling a Cored Hole

Castings having cored holes are usually drilled with a four lip drill. The hole in the casting should be beveled, as shown in Fig. 195, to start the drill true; otherwise, the drill will follow the cored hole and may be thrown off center. For accurate drilling it is advisable to counterbore the hole a short depth to give the drill point a perfectly concentric starting point.

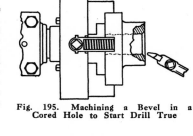

Fig. 195. Machining a Bevel in a Cored Hole to Start Drill True

How to Sharpen Drills

Correct grinding of the drill point is essential for accuracy and efficiency in all drilling operations. A medium grain grinding wheel that has been dressed true should be used for grinding drill points. The drill point should not be overheated by grinding or the temper may be drawn.

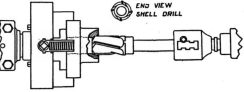

Fig. 196. Drilling a Cored Hole with a Four Lip Shell Drill

Before grinding a drill, study the point of a new drill as received from the manufacturer; then try to duplicate it. This can be accomplished by holding the drill at the correct angle with the grinding wheel

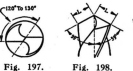

Fig. 197. Correct Point Angle

Fig. 198. Correct Lip Angle

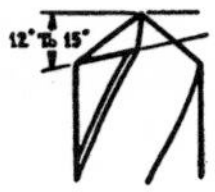

Fig. 199. Correct Clearance

and giving the drill point a wiping motion as it is ground, lowering the shank end of the drill and giving the drill a slight twist to the right simultaneously. It is very important that both lips of the drill be ground exactly the same.

The angle of the chisel point or dead center should be from 120° to 130°, as shown in Fig. 197. The cutting lips "L" Fig. 198, should be exactly the same length and angle; otherwise the drill will cut oversize. The best angle for general work is 59°, as indicated.

The clearance back of the cutting edge should be from 12° to 15°, as shown in Fig. 199. Less clearance may prevent the drill from cutting freely, and more clearance will cause the cutting edge to dull quickly.

A drill grinding gauge similar to the one shown in Fig. 200 will aid in grinding the correct angle and length of cutting lip on the drill point.

Fig. 200. Drill Grinding Gauge

Reaming in the Lathe

Reamers are used in the lathe to finish a number of holes quickly and accurately to the same diameter. Usually the hole is first drilled or bored roughly to size, allowing sufficient stock for reaming. Two types of reamers are used, the rose reamer and the fluted reamer.

Rose reamers are ground for cutting on the end only and are intended for rough reaming as they do not produce a good finish or an accurate diameter.

Fluted reamers are ground for cutting on both the ends and the sides of the blades and are usually used after the rose reamer to obtain an exact size and produce a good smooth finish. Fluted reamers should be used only for light cuts, removing not over .010 in. from the hole.

Reamer in Drill Chuck

Straight shank reamers are usually held in a drill chuck, as shown in Fig. 201. Taper shank reamers may be inserted direct in the tailstock spindle. The reamer is fed carefully through the hole by turning the tailstock handwheel. Always use a slow spindle speed and when reaming steel keep the reamer flooded with lard oil.

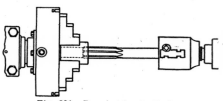

Fig. 201. Reaming in the Lathe

Floating Reamer Driver

For some reaming operations it is desirable for the reamer to follow a bored hole as accurately as possible, and for this type of work a floating reamer driver similar to the one shown in Fig. 202 is used.

Large reamers are sometimes supported on the tailstock center point. A lathe dog is attached to the reamer shank and a stick with one end resting against the lathe bed is placed between the reamer shank and the tail of the dog.

Fig. 202. Floating Reamer Driver

Tapping Threads

Threads may be tapped in the lathe, using a tap as shown in Fig. 203. The lathe spindle should be operated at slow speed and the tap fed to the work by turning the tailstock handwheel, or by sliding the entire tailstock on the lathe bed. Taps may also be held in a drill chuck.

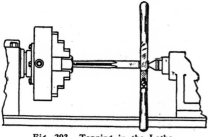

Fig. 203. Tapping in the Lathe

Fig. 204. Cutting Screw Threads in the Lathe

Chapter X

CUTTING SCREW THREADS

Cutting screw threads in the lathe is accomplished by connecting the headstock spindle of the lathe with the lead screw by a series of gears so that a positive carriage feed is obtained and the lead screw is driven at the required speed with relation to the headstock spindle.

The gearing between the headstock spindle and lead screw may be arranged so that any desired pitch of the thread may be cut. For example, if the lead screw has eight threads per inch and the gears are arranged so that the headstock spindle revolves four times while the lead screw revolves once, the thread cut will be four times as fine as the thread on the lead screw or 32 threads per inch.

The cutting tool is ground to the shape required for the form of the thread to be cut, that is, American National Form, "V," Acme, Square, Whitworth, International Metric, etc.

Either right hand or left hand threads may be cut by reversing the direction of rotation of the lead screw. This may be accomplished by shifting the reverse lever on the headstock.

Fig. 205. Acme Screw Thread

Fig. 206. National Coarse Thread

Fig. 207. Double Square Thread

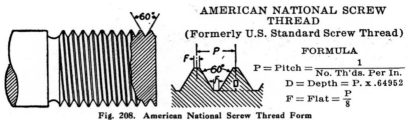

AMERICAN NATIONAL SCREW
THREAD
(Formerly U.S. Standard Screw Thread)

FORMULA

$$P = \text{Pitch} = \frac{1}{\text{No. Th'ds. Per In.}}$$
$$D = \text{Depth} = P. \text{ x } .64952$$
$$F = \text{Flat} = \frac{P}{8}$$

Fig. 208. American National Screw Thread Form

American National Screw Threads

The National Screw Thread Commission in 1928 was authorized by Congress to establish a standard system of screw threads for use in the United States. As a result this commission established the American National Screw Thread System which has been approved by the Secretary of War, Secretary of the Navy, and Congress, and is now generally used by all shops in the United States.

The form of the thread adopted is shown above and tables for both the Fine Thread Series and Coarse Thread Series are given on page 71. A report of the National Screw Thread Commission defines the following terms.

Terms Relating to Screw Threads

Screw Thread. A ridge of uniform section in the form of a helix on the surface of a cylinder or cone.

External and Internal Threads. An external thread is a thread on the outside of a member. Example: A threaded plug. An internal thread is a thread on the inside of a member. Example: A threaded hole.

Major Diameter (formerly known as "outside diameter"). The largest diameter of the thread of the screw or nut. The term "major diameter" replaces the term "outside diameter" as applied to the thread of a screw and also the term "full diameter" as applied to the thread of a nut.

Minor Diameter (formerly known as "core diameter"). The smallest diameter of the thread of the screw or nut. The term "minor diameter" replaces the term "core diameter" as applied to the thread of a screw and also the term "inside diameter" as applied to the thread of a nut.

Pitch Diameter. On a straight screw thread, the diameter of an imaginary cylinder, the surface of which would pass through the threads at such points as to make equal the width of the threads and the width of the spaces cut by the surface of the cylinder.

Pitch. The distance from a point on a screw thread to a corresponding point on the next thread measured parallel to the axis.

Lead. The distance a screw thread advances axially in one turn. On a single-thread screw, the lead and pitch are identical; on a double-thread screw the lead is twice the pitch; on a triple-thread screw, the lead is three times the pitch, etc.

TABLES OF AMERICAN NATIONAL STANDARD SCREW THREAD PITCHES AND RECOMMENDED TAP DRILL SIZES

American National Coarse Standard Thread (N.C.)
Formerly U. S. Standard

Sizes	Threads Per Inch	Outside Diameter of Screw	Tap Drill Sizes	Decimal Equivalent of Drill
1	64	.073	53	0.0595
2	56	.086	50	0.0700
3	48	.099	47	0.0785
4	40	.112	43	0.0890
5	40	.125	38	0.1015
6	32	.138	36	0.1065
8	32	.164	29	0.1360
10	24	.190	25	0.1495
12	24	.216	16	0.1770
¼	20	.250	7	0.2010
5/16	18	.3125	F	0.2570
⅜	16	.375	5/16	0.3125
7/16	14	.4375	U	0.3680
½	13	.500	27/64	0.4219
9/16	12	.5625	31/64	0.4843
⅝	11	.625	17/32	0.5312
¾	10	.750	21/32	0.6562
⅞	9	.875	49/64	0.7656
1	8	1.000	⅞	0.875
1⅛	7	1.125	63/64	0.9843
1¼	7	1.250	1 7/64	1.1093

American National Fine Standard Thread (N.F.)
Formerly S. A. E. Thread

Sizes	Threads Per Inch	Outside Diameter of Screw	Tap Drill Sizes	Decimal Equivalent of Drill
0	80	.060	3/64	0.0469
1	72	.073	53	0.0595
2	64	.086	50	0.0700
3	56	.099	45	0.0820
4	48	.112	42	0.0935
5	44	.125	37	0.1040
6	40	.138	33	0.1130
8	36	.164	29	0.1360
10	32	.190	21	0.1590
12	28	.216	14	0.1820
¼	28	.250	3	0.2130
5/16	24	.3125	I	0.2720
⅜	24	.375	Q	0.3320
7/16	20	.4375	25/64	0.3906
½	20	.500	29/64	0.4531
9/16	18	.5625	0.5062	0.5062
⅝	18	.625	0.5687	0.5687
¾	16	.750	11/16	0.6875
⅞	14	.875	0.8020	0.8020
1	14	1.000	0.9274	0.9274
1⅛	12	1.125	1 3/64	1.0468
1¼	12	1.250	1 11/64	1.1718

TABLES OF AMERICAN NATIONAL SPECIAL SCREW THREAD PITCHES (N.S.) AND RECOMMENDED TAP DRILL SIZES

Sizes	Threads Per Inch	Outside Diameter of Screw	Tap Drill Sizes	Decimal Equivalent of Drill
¼	24	.250	4	0.2090
	27		3	0.2130
	32		7/32	0.2187
5/16	20	.3125	17/64	0.2656
	27		J	0.2770
	32		9/32	0.2812
⅜	20	.375	21/64	0.3281
	27		R	0.3390
7/16	24	.4375	X	0.3970
	27		Y	0.4040

Sizes	Threads Per Inch	Outside Diameter of Screw	Tap Drill Sizes	Decimal Equivalent of Drill
½	12	.500	27/64	0.4219
	24		29/64	0.4531
	27		15/32	0.4687
9/16	27	.5625	17/32	0.5312
⅝	12	.625	35/64	0.5469
	27		19/32	0.5937
¾	12	.750	43/64	0.6719
	27		23/32	0.7187
⅞	12	.875	51/64	0.7969
	18		53/64	0.8281
	27		27/32	0.8437
1	12	1.000	59/64	0.9219
	27		31/32	0.9687

Fig. 209.   Standard Change Gear Lathe Set Up for Cutting Screw Threads

Cutting Threads on Standard Change Gear Lathes

Screw threads are cut on Standard Change Gear Lathes by engaging the apron half nuts with the lead screw. The pitch of thread to be cut is determined by the number of teeth in the change gears used on the reverse stud and the lead screw, also the compound gears used.

To set up the lathe for cutting a screw thread, first determine the number of threads per inch to be cut. By referring to the change gear chart attached to the lathe (Fig. 210) the change gears required can be determined. The thread to be cut should be located in the first column under the heading "Threads Per Inch." In the second column under the heading "Stud Gear" is listed the number of teeth in the change gear which should be placed on the reverse stud "A" of the lathe. (See Fig. 209.) In the third column under the heading "Idler Gear" is listed the figure number on the index chart showing the arrangement of idler gear "B" and compound gears. In the fourth column under the heading "Screw Gear" is listed the number of teeth in the gear to be placed on the lead screw "C".

After selecting the change gears necessary for cutting the desired thread, place them on the reverse stud and lead screw respectively and connect them with the idler gear and compound gears, as shown on the change gear chart.

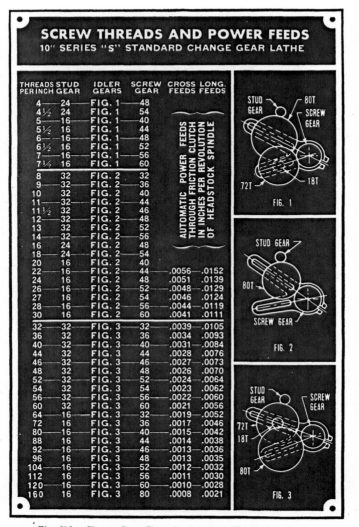

SCREW THREADS AND POWER FEEDS
10" SERIES "S" STANDARD CHANGE GEAR LATHE

THREADS PER INCH	STUD GEAR	IDLER GEARS	SCREW GEAR	CROSS FEEDS	LONG. FEEDS
4	24	FIG. 1	48		
4½	24	FIG. 1	54		
5	16	FIG. 1	40		
5½	16	FIG. 1	44		
6	16	FIG. 1	48		
6½	16	FIG. 1	52		
7	16	FIG. 1	56		
7½	16	FIG. 1	60		
8	32	FIG. 2	32		
9	32	FIG. 2	36		
10	32	FIG. 2	40		
11	32	FIG. 2	44		
11½	32	FIG. 2	46		
12	32	FIG. 2	48		
13	32	FIG. 2	52		
14	32	FIG. 2	56		
16	24	FIG. 2	48		
18	24	FIG. 2	54		
20	16	FIG. 2	40		
22	16	FIG. 2	44	.0056	.0152
24	16	FIG. 2	48	.0051	.0139
26	16	FIG. 2	52	.0048	.0129
27	16	FIG. 2	54	.0046	.0124
28	16	FIG. 2	56	.0044	.0119
30	16	FIG. 2	60	.0041	.0111
32	32	FIG. 3	32	.0039	.0105
36	32	FIG. 3	36	.0034	.0093
40	32	FIG. 3	40	.0031	.0084
44	32	FIG. 3	44	.0028	.0076
46	32	FIG. 3	46	.0027	.0073
48	32	FIG. 3	48	.0026	.0070
52	32	FIG. 3	52	.0024	.0064
54	32	FIG. 3	54	.0023	.0062
56	32	FIG. 3	56	.0022	.0060
60	32	FIG. 3	60	.0021	.0056
64	16	FIG. 3	32	.0019	.0052
72	16	FIG. 3	36	.0017	.0046
80	16	FIG. 3	40	.0015	.0042
88	16	FIG. 3	44	.0014	.0038
92	16	FIG. 3	46	.0013	.0036
96	16	FIG. 3	48	.0013	.0035
104	16	FIG. 3	52	.0012	.0032
112	16	FIG. 3	56	.0011	.0030
120	16	FIG. 3	60	.0010	.0028
160	16	FIG. 3	80	.0008	.0021

AUTOMATIC POWER FEEDS THROUGH FRICTION CLUTCH IN INCHES PER REVOLUTION OF HEADSTOCK SPINDLE

FIG. 1 — STUD GEAR, 80T, SCREW GEAR, 72T, 18T
FIG. 2 — STUD GEAR, 80T, SCREW GEAR
FIG. 3 — STUD GEAR, SCREW GEAR, 72T, 18T, 80T

Fig. 210. Change Gear Chart for Standard Change Gear Lathes

Change Gear Chart

A thread cutting chart similar to the one above in Fig. 210 showing the gearing required for various pitches of screw threads and various power turning feeds is attached to each Standard Change Gear Lathe.

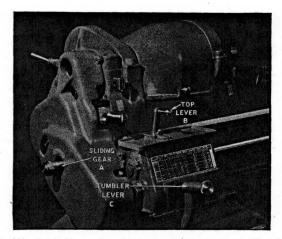

Fig. 211. Quick Change Gear Mechanism for Cutting Screw Threads

Cutting Screw Threads on Quick Change Gear Lathes

The Quick Change Gear Lathe is fitted with a gear box, shown in Fig. 211, which permits obtaining various pitches of screw threads without the use of loose change gears. The screw thread chart attached to the gear box is shown in Fig. 212 below. This chart reads directly in threads per inch. It is only necessary to arrange the levers of the gear box as indicated on the index plate in order to obtain various screw threads and feeds.

The pitch of the thread to be cut is determined by shifting the sliding gear A, top lever B and tumbler lever C of the quick change gear box so that they conform with the thread cutting chart. For example, to cut six threads per inch the sliding gear A is pushed in, the top lever B on the gear box is pushed to the extreme left position and the tumbler lever C is placed just below the column in which the thread appears on the index chart.

SLIDING GEAR	TOP LEVER	|THREADS PER INCH—FEEDS IN THOUSANDTHS								AUTOMATIC CROSS FEED EQUALS .375 TIMES LONGITUDINAL FEED
IN	LEFT	4 .0841	4½ .0748	5 .0673	5½ .0612	5¾ .0585	6 .0561	6½ .0518	7 .0481	
	CENTER	8 .0421	9 .0374	10 .0337	11 .0306	11½ .0293	12 .0280	13 .0259	14 .0240	
	RIGHT	16 .0210	18 .0187	20 .0168	22 .0153	23 .0146	24 .0140	26 .0129	28 .0120	
OUT	LEFT	32 .0105	36 .0093	40 .0084	44 .0076	46 .0073	48 .0070	52 .0065	56 .0060	
	CENTER	64 .0053	72 .0047	80 .0042	88 .0038	92 .0037	96 .0035	104 .0032	112 .0030	
	RIGHT	128 .0026	144 .0023	160 .0021	176 .0019	184 .0018	192 .0017	208 .0016	224 .0015	

16-INCH SOUTH BEND QUICK CHANGE GEAR LATHE

Fig. 212. Index Plate for Threads and Feeds on Quick Change Gear Lathe

Tools for Cutting Screw Threads

The shape or form of a screw thread cut on the lathe is determined by the shape of the cutter bit, which must be carefully ground and set if an accurate thread form is to be obtained. The most common thread forms are shown on pages 70, 82, 83 and 84. A gauge should be used for grinding the lathe tool to the required shape for any form of screw thread.

Use of Center Gauge

The point of the cutter bit must be ground to an angle of 60° for cutting American National screw threads in the lathe, as shown in Fig. 213 at right. A center gauge having a 60° included angle is used for grinding the tool to the exact angle required. The top of

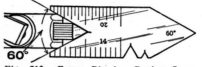
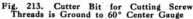

Fig. 213. Cutter Bit for Cutting Screw Threads is Ground to 60° Center Gauge

the tool is usually ground flat, with no side rake or back rake. However, for cutting threads in steel, side rake is sometimes used.

Front Clearance

There must be sufficient front clearance on the cutter bit to permit it to cut freely. Usually the front clearance is sufficient to prevent the tool from dragging in the helix angle of the thread so that except for very coarse pitches the helix angle may be ignored.

Fig. 214. Side View of Lathe Tool Cutter Bit Ground for Cutting Screw Threads

Formed Threading Tool

A formed threading tool is sometimes used if considerable threading is to be done. Fig. 215 illustrates a good type of formed threading tool. The formed threading tool requires grinding on top only to sharpen, and therefore always remains true to form and correct angle.

Fig. 215. Formed Thread Cutting Tool, Solid Type

Thread Tool Gauge

A gauge for grinding threading tools to the exact shape required for various pitches of American National screw threads is shown in Fig. 216.

For American National Screw Threads finer than 10 per inch, the point of the tool is usually left sharp or with a very small flat. However, for coarser pitches of threads and when maximum strength is desired, the flat on the point of the tool should be one-eighth of the pitch. (See Fig. 208, page 70.)

Fig. 216. Standard Screw Thread Tool Gauge for Grinding Thread Cutting Tools

Setting Cutter Bit for External Threads

The top of the threading tool should be placed exactly on center, as shown in Fig. 217 at right, for cutting external screw threads. Note that the top of the tool is ground flat and is in exact alignment with the lathe center. This is necessary to obtain the correct angle of the thread.

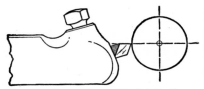

Fig. 217. Top of Cutter Bit Set On Center for Cutting Screw Threads

The threading tool must be set square with the work, as shown in Fig. 218. The center gage is used to adjust the point of the threading tool, and if the tool is carefully set a perfect thread will result. Of course, if the threading tool is not set perfectly square with the work, the angle of the thread will be incorrect.

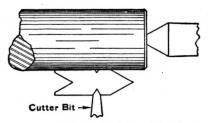

Cutter Bit →

Fig. 218. Cutter Bit Set Square With Work, for Cutting External Screw Threads

Setting Cutter Bit for Internal Threads

The point of the threading tool is also placed exactly on center, as shown in Fig. 219 at right for cutting internal screw threads. The point of the tool must be set perfectly square with the work. This may be accomplished by fitting the point of the tool into the center gauge, as shown in Fig. 220 at right.

Fig. 219. For Cutting Internal Screw Threads, Top of Cutter Bit Should be Set Exactly on Center

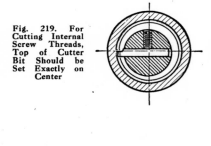

When adjusting the threading tool for cutting internal threads, allow sufficient clearance between the tool and the inside diameter of the hole to permit backing out the tool when the end of the cut has been reached. However, the boring bar should be as large in diameter and as short as possible to prevent springing.

When cutting internal screw threads more front clearance is required to prevent the heel of the tool from rubbing than when cutting external threads.

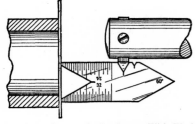

Fig. 220. Cutter Bit Set Square With Work for Cutting Internal Screw Threads

Fig. 221. Compound Rest Set at 29° Angle for Cutting Screw Threads

Position of Compound Rest for Cutting Screw Threads

In manufacturing plants where maximum production is desired, it is customary to place the compound rest of the lathe at an angle of 29° for cutting screw threads.

The compound rest is swung around to the right, as shown in Figs. 221 and 224. The compound rest screw is used for adjusting the depth of cut and most of the metal is removed by the left side of the threading tool. (See Fig. 223.) This permits the chip to curl out of the way better than if tool is fed straight in.

The right side of the tool will shave the thread smooth and produce a good finish, although it does not remove enough metal to interfere with the main chip which is taken by the left side of the tool.

Fig. 222. Cutting a Screw Thread with Compound Rest Set at 29°

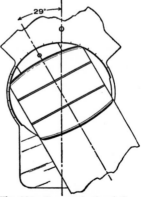

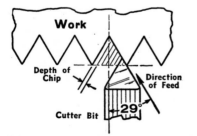

Fig. 223. Action of Thread Cutting Tool when Compound Rest is Set at 29° Angle

Fig. 224. Correct Angle of Compound Rest for Thread Cutting

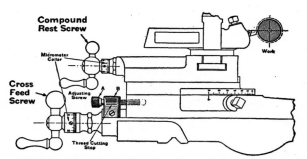

Fig. 225. Thread Cutting Stop Attached to Dovetail of Saddle

Use of Thread Cutting Stop

On account of the lost motion caused by the play necessary for smooth operation of the change gears, lead screw, half nuts, etc., the thread cutting tool must be withdrawn quickly at the end of each cut, before the lathe spindle is reversed to return the tool to the starting point. If this is not done, the point of the tool will dig into the thread and may be broken off. The thread cutting stop may be used for regulating the depth of each successive chip.

The point of the tool should first be set so that it just touches the work; then lock the thread cutting stop and turn the thread cutting stop screw until the shoulder is tight against the stop. When ready to take the first chip run the tool rest back by turning the cross feed screw to the left several turns and move the tool to the point where the thread is to start. Then turn the cross feed screw to the right until the thread cutting stop screw strikes the thread cutting stop. The tool rest is now in the original position, and by turning the compound rest feed screw in .002 in. or .003 in. the tool will be in a position to take the first cut.

Using Micrometer Collar

The micrometer collar on the cross feed screw of the lathe may be used in place of the thread cutting stop, if desired. To do this, first bring the point of threading tool up so that it just touches the work, then loosen small set screw "A," adjust the micrometer collar on the cross feed screw to zero, and tighten set screw.

All adjusting for obtaining the desired depth of cut should be done with the compound rest screw. Withdraw the tool at the end of each cut by turning the cross feed screw to the left one complete turn, return the tool to the starting point and turn the cross feed screw to the right one turn, stopping at zero. The compound rest feed screw may then be adjusted for any desired depth of chip.

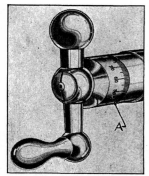

Fig. 226. Micrometer Collar on Cross Feed Screw of Lathe

Taking the First Cut

After setting up the lathe, as explained on the preceding pages, take a very light trial cut just deep enough to scribe a line on the surface of the work, as shown in Fig. 227. The purpose of this trial cut is to make sure the lathe is arranged for cutting the desired pitch of thread.

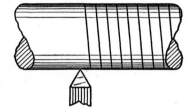

Fig. 227. Trial Cut to Check the Set-up for Thread Cutting

Measuring Screw Threads

To check the number of threads per inch, place a scale against the work, as shown in Fig. 228, so that the end of the scale rests on the point of the thread or on one of the scribed lines. Count the spaces between the end of the scale and the first inch mark, and this will give you the number of threads per inch. Fig. 228 shows eight threads per inch.

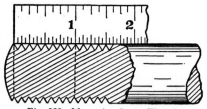

Fig. 228. Measuring Screw Threads

Screw Pitch Gauge

A screw thread gauge as illustrated in Fig. 229 is very convenient for checking the finer pitches of screw threads. This gauge consists of a number of sheet metal plates in which are cut the exact form of the various pitches of threads.

Fitting and Testing Threads

The final check for both the diameter and pitch of the thread may be made with the nut that is to be used or with a ring thread gauge, if one is available. Fig. 230 shows how the nut may be used for checking the thread. The nut should fit snugly without play or shake but should not bind on the thread at any point.

If the angle of the thread is correct and the thread is cut to the correct depth it will fit the nut perfectly. However, if the angle of the thread is incorrect, the thread may appear to fit the nut but will only be touching at a few points. For this reason, the thread should be checked by other methods in addition to the nut or ring gauge.

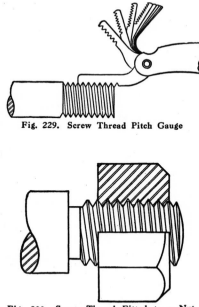

Fig. 229. Screw Thread Pitch Gauge

Fig. 230. Screw Thread Fitted to a Nut

Resetting Tool After Cut Has Been Started

If for any reason it is necessary to remove the thread cutting tool before the thread has been completed, the tool must be carefully readjusted so that it will follow the original groove when it is replaced in the lathe.

Before adjusting the tool, take up all the lost motion by pulling the belt forward by hand.

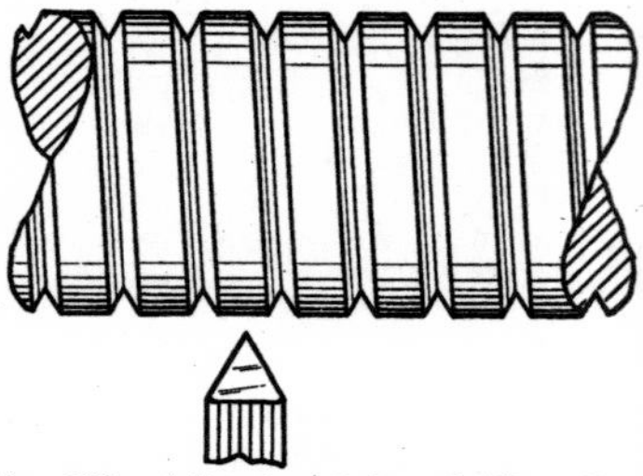

Fig. 231. Adjusting Point of Threading Tool to Conform with Thread

The compound rest top should be set at an angle, and by adjusting the cross feed screw and compound rest feed screw simultaneously the point of the tool can be made to enter exactly into the original groove.

Finishing the End of a Thread

The end of the thread may be finished by any one of several methods. The 45° chamfer on the end of the thread, as shown in Fig. 232, is commonly used for bolts, cap screws, etc. For machine parts and special screws the end is often finished by rounding with a forming tool, as shown in Fig. 233.

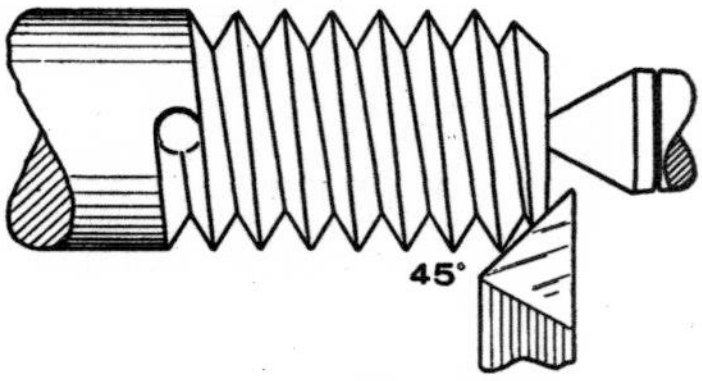

Fig. 232. Finish End of Thread with 45° Chamfer

It is difficult to stop the threading tool abruptly so some provision is usually made for clearance at the end of the cut. In Fig. 232 a hole has been drilled in the end of the shaft, and in Fig. 233 a neck or groove has been cut around the shaft. The groove is preferable as the lathe must be run very slowly in order to obtain satisfactory results with the drilled hole.

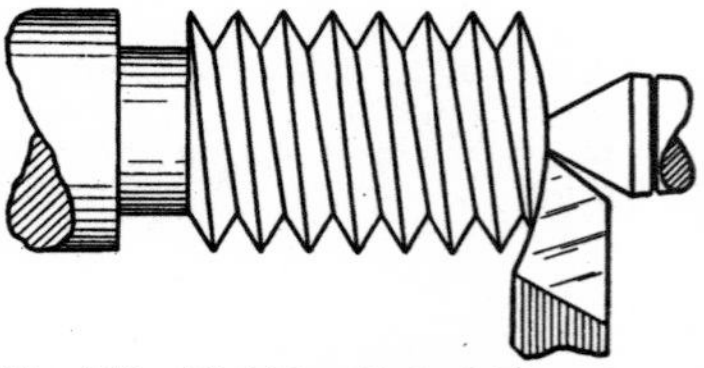

Fig. 233. Finishing End of Thread with Forming Tool

Cutting a Left Hand Screw Thread

A left hand screw is one that turns counter-clockwise when advancing (looking at head of screw) as shown in Fig. 235. This is just the opposite of a right hand screw. Left hand threads are used for the cross feed screws of lathes, the left hand end of axles for automobiles and wagons, one end of a turnbuckle, some pipe threads, etc.

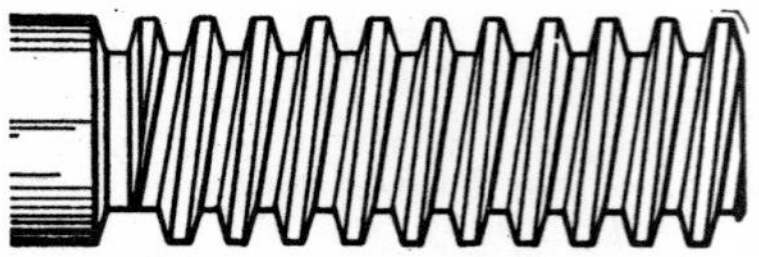

Fig. 234. A Left Hand Screw Thread

In cutting left hand screw threads the lathe is set up exactly the same as for cutting right hand screw threads, except that the lathe must be arranged to feed the tool from left to right, instead of from right to left, when the spindle is revolving forward.

Fig. 235. A Left Hand Screw Advances When Turned Counter-clockwise

Fig. 236.   Thread Dial Indicator Attached to
Lathe Carriage

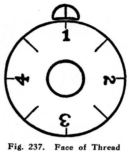

Fig. 237.   Face of Thread
Dial Indicator

Use of Thread Dial Indicator

A thread dial indicator is usually used for cutting long screw threads. This device permits disengaging the half nuts at the end of the cut, returning the carriage to the starting point by hand, and then engaging the half nuts at the correct time so that the tool will follow the original cut. As the lead screw revolves the dial is turned and the numbers on the dial indicate points at which the half nuts may be engaged, which are as follows:

For all even numbered threads, close the half nuts at any line on the dial.

For all odd numbered threads, close the half nuts at any numbered line on dial.

For all threads involving one-half of a thread in each inch, such as 11½, close the half nuts at any odd numbered line.

For quarter threads or eighth threads, return to the original starting point before closing the half nuts.

The thread dial cannot be used with transposing gears for cutting metric threads.

Use Oil When Cutting Threads in Steel

Lard oil (or machine oil) should be used when cutting screw threads in steel in order to produce a smooth thread. If oil is not used, a very rough finish will be caused by tearing of the steel by the cutting tool.

The oil should be applied generously preceding each cut. A small paint brush is ideal for applying the oil when cutting external screw threads, as illustrated in Fig. 238.

Fig. 238.   Using a Small Brush to Apply
Oil When Cutting Screw Threads

Tapered Screw Threads

Tapered screw threads such as pipe threads may be cut with the aid of a taper attachment, as shown in Fig. 239, or by setting over the tailstock off center, as shown in Fig. 240.

Regardless of which method is used, it is important that the thread tool is set square with the straight portion of the work, as shown in Figs. 239 and 240 and not to the tapered portion. The angle of the sides of the thread will be incorrect if the tool is not set as shown.

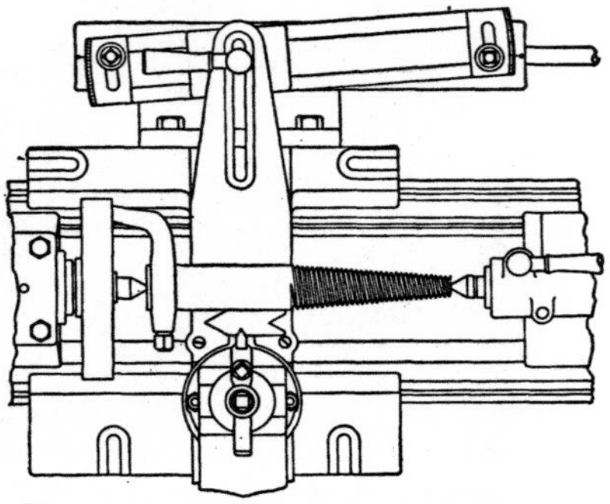

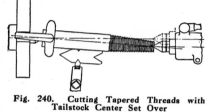

Fig. 239. Cutting Tapered Threads with Taper Attachment

Square Screw Threads

Square threads are used for vice screws, jack screws, etc. The sides of the tool for cutting square threads should be ground at an angle to conform with the helix angle of the thread, as shown in Fig. 241.

Fig. 240. Cutting Tapered Threads with Tailstock Center Set Over

To determine the helix angle of a screw thread, draw line A-C2 equal to the circumference of the thread to be cut. Draw line C2-C equal to the lead of the thread and at right angles to line A-C2. Complete the triangle by drawing line A-C. Angle B in the triangle is the helix angle of the thread. The sides of the tool E and F should be given a little clearance.

The width of the cutting edge of the tool for cutting square screw threads is exactly one-half the pitch, but the width of the tool for threading the nut should be from one thousandth to three thousandths of an inch larger, to permit a free fit on the screw.

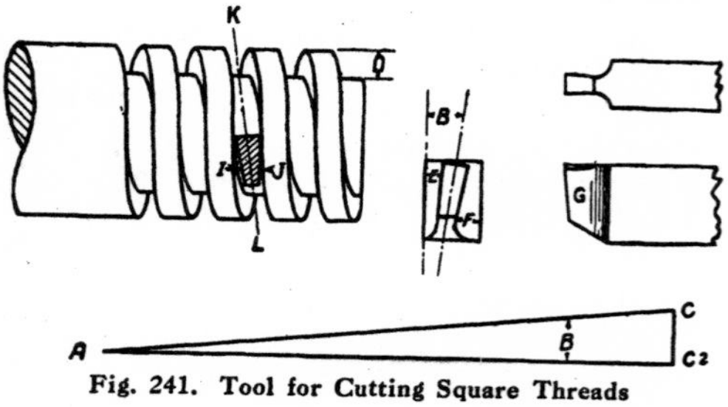

Fig. 241. Tool for Cutting Square Threads

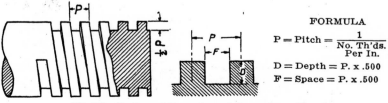

Fig. 242. Design and Proportions of Square Screw Threads

FORMULA

$$P = \text{Pitch} = \frac{1}{\substack{\text{No. Th'ds.} \\ \text{Per In.}}}$$

$$D = \text{Depth} = P. \times .500$$

$$F = \text{Space} = P. \times .500$$

Acme Screw Threads

FORMULA

P = Pitch = $\dfrac{1}{\text{No. Th'ds. Per In.}}$

D = Depth = ½ P. + .010 In.

F = Flat = .3707 P.

C = Flat = .3707 P. — .0052 In.

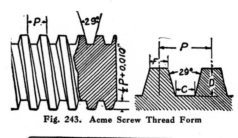

Fig. 243. Acme Screw Thread Form

Acme screw threads are used for the feed screws and adjusting screws of machine tools and machinery of all kinds. Acme threads are preferable to square threads because they are easier to cut.

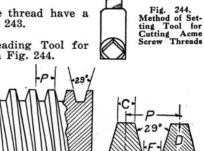

Fig. 244. Method of Setting Tool for Cutting Acme Screw Threads

While the top and the bottom of the threads are similar to a square thread in that they are flat, the sides of the thread have a 29° included angle, as shown in Fig. 243.

The method of setting a Threading Tool for cutting an Acme Thread is shown in Fig. 244.

29° Worm Thread
(Brown & Sharpe)

FORMULA

P = Pitch = $\dfrac{1}{\text{No. Th'ds. Per In.}}$

D = Depth = .6866 P.

F = Flat = .31 P.

C = Flat = .335 P.

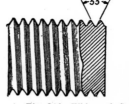

Fig. 245. 29° Worm Thread Form

A 29° Worm Thread should not be confused with the Acme Standard Thread because it differs in the depth of the thread, the width of the top of the tooth and the width of the bottom of the tooth, as shown above.

Whitworth Thread

FORMULA

P = Pitch = $\dfrac{1}{\text{No. Th'ds. Per In.}}$

D = Depth = P. x .6403

R. = Radius = .1373 P. $\dfrac{1}{\text{No. Th'ds. Per In.}}$

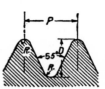

Fig. 246. Whitworth Screw Thread Form

The Whitworth thread form is used mostly in England. Great care is required in grinding the thread tool so that it will produce the radius at the top and bottom of the thread.

Metric and English Transposing Gears

When it is desired to cut both English and metric screw threads on the same lathe, transposing gears are required.

English transposing gears are used for cutting English screw threads on lathes having metric lead screws. Metric transposing gears are used for cutting metric screw threads on lathes having English lead screws.

The form of the metric thread is similar to the American National Screw Thread form, having a 60° included angle and a flat at the top of the thread, but a small radius at the root of the thread provides greater clearance. (See Fig. 250.)

Fig. 247. Lathe Equipped with Transposing Gears

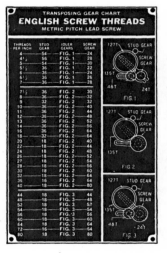

TRANSPOSING GEAR CHART
ENGLISH SCREW THREADS
METRIC PITCH LEAD SCREW

THREADS PER INCH	STUD GEAR	IDLER GEARS	SCREW GEAR
4	54	FIG. 1	24
4½	56	FIG. 1	28
5	54	FIG. 1	30
5½	36	FIG. 1	22
6	36	FIG. 1	24
6½	36	FIG. 1	26
7	36	FIG. 1	28
7½	36	FIG. 2	30
8	36	FIG. 2	32
9	32	FIG. 2	32
10	36	FIG. 2	40
11	36	FIG. 2	44
12	36	FIG. 2	48
14	36	FIG. 2	56
16	36	FIG. 2	64
18	32	FIG. 2	64
20	18	FIG. 2	40
22	18	FIG. 2	44
26	18	FIG. 2	52
27	18	FIG. 2	54
28	18	FIG. 2	56
30	18	FIG. 2	60
32	18	FIG. 2	64
36	18	FIG. 2	72
40	18	FIG. 2	80
44	18	FIG. 3	44
48	18	FIG. 3	48
52	18	FIG. 3	52
54	18	FIG. 3	54
56	18	FIG. 3	56
60	18	FIG. 3	60
64	16	FIG. 3	64
72	16	FIG. 3	72
80	18	FIG. 3	80

Left—Fig. 248 Index Chart Showing English Threads Cut on Metric Lathe with English Transposing Gears

Right — Fig. 249 Index Chart Showing Metric Threads Cut on English Lathe with Metric Transposing Gears

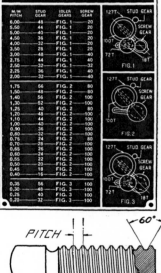

TRANSPOSING GEAR CHART
METRIC SCREW THREADS
ENGLISH PITCH LEAD SCREW

M/M PITCH	STUD GEAR	IDLER GEARS	SCREW GEAR
6.00	48	FIG. 1	20
5.50	44	FIG. 1	20
5.00	40	FIG. 1	20
4.50	36	FIG. 1	20
4.00	32	FIG. 1	20
3.50	28	FIG. 1	20
3.00	48	FIG. 1	40
2.75	44	FIG. 1	40
2.50	32	FIG. 1	32
2.25	36	FIG. 1	40
2.00	32	FIG. 1	40
1.75	56	FIG. 2	80
1.50	48	FIG. 2	80
1.40	56	FIG. 2	100
1.30	52	FIG. 2	100
1.25	40	FIG. 2	80
1.20	48	FIG. 2	100
1.10	44	FIG. 2	100
1.00	40	FIG. 2	100
0.90	36	FIG. 2	100
0.80	32	FIG. 2	100
0.75	24	FIG. 2	80
0.70	28	FIG. 2	100
0.65	26	FIG. 2	100
0.60	24	FIG. 2	100
0.55	22	FIG. 2	100
0.50	20	FIG. 2	100
0.45	18	FIG. 2	100
0.40	16	FIG. 2	100
0.35	56	FIG. 3	100
0.30	48	FIG. 3	100
0.25	40	FIG. 3	100
0.20	32	FIG. 3	100

FORMULA

P = Pitch in MM
D = Depth of Engagement = P x .6495
H = Depth of Thread = P x .6945
r = Maximum Radius = P x .0633
$F = Flat = \dfrac{P}{8}$

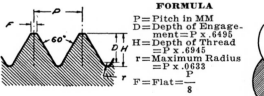

Fig. 250. International Standard Metric Screw Thread Form

Fig. 251. 2.5 mm. Pitch Metric Screw Thread

Fig. 252. A Metric Quick Change Gear Lathe

Metric Lathe with Metric Lead Screw

Metric lathes equipped with metric lead screws are preferable in locations where metric screw threads are used exclusively. The metric lathe is identical with the English lathe, except that the lead screw, cross feed screw and compound rest screw have metric threads, and all graduations are in the metric system.

Metric lathes are made in both the Standard Change Gear and Quick Change Gear types. Metric Quick Change Gear Lathes have a quick change gear box which permits cutting a wide range of metric screw threads and feeds, as listed on the index chart, which is illustrated below in Fig. 253. Metric Standard Change Gear Lathes have a similar range of metric screw threads and feeds.

MANUFACTURED BY SOUTH BEND LATHE WORKS SOUTH BEND, IND., U.S.A.										
PITCHES IN mm—PASOS EN mm—PAS EN mm								POSITION POSICION	STUD ARBOL ARBRE	
7.500	7.000	6.500	6.000	5.500	5.000	4.500	4.000	D	50	
3.750	3.500	3.250	3.000	2.750	2.500	2.250	2.000	C	"	
1.875	1.750	1.625	1.500	1.375	1.250	1.125	1.000	B	"	
1.500	1.400	1.300	1.200	1.100	1.000	0.900	0.800	C	20	
0.750	0.700	0.650	0.600	0.550	0.500	0.450	0.400	B	"	
0.375	0.350	0.325	0.300	0.275	0.250	0.225	0.200	A	"	
FEEDS IN mm—AVANCES EN mm										
0.512	0.478	0.444	0.410	0.375	0.341	0.307	0.273	C	20	
0.256	0.239	0.222	0.205	0.188	0.171	0.154	0.137	B	"	
0.128	0.119	0.111	0.102	0.094	0.085	0.077	0.068	A	"	

9-Inch—235 mm SOUTH BEND LATHE MODEL A
CATALOG NO.
BED LENGTH
PAT. APP. FOR

Positions / Posiciones   A   B   C   D

Fig. 253. Index Chart Showing Metric Threads and Feeds on a 9-inch Swing Metric Quick Change Gear Lathe

Cutting Multiple Screw Threads

A multiple thread having two grooves is known as a double thread, three threads a triple thread, etc. (See Fig. 255.) The pitch and lead of a multiple thread should not be confused. The pitch is the distance from a point on one thread to the corresponding point on the next thread, while the lead is the distance a screw thread advances in one turn.

Fig. 255. A Multiple Screw Thread Having Two Grooves (Double Thread)

When cutting multiple threads in the lathe the first thread is cut to the desired depth. The work is then revolved part of a turn, and the second thread cut, etc. In order to obtain an exact spacing it is advisable to mill as many equally spaced slots in the face plate for the lathe dog as there are multiple threads to be cut. For a double thread, two slots; a triple thread, three slots, etc. If it is not convenient to cut slots in the face plate, equally spaced studs may be attached to the face plate and a straight tail lathe dog used.

Another method for indexing the work when cutting multiple threads is to disengage the change gears after the first thread has been completed and turn the spindle to the required position for starting the next cut.

Cutting Threads with Die in Tailstock

A die may be mounted on tailstock of lathe for cutting screw threads, as shown in Fig. 256. A lathe tool may also be mounted in the tool post for turning operations or for cutting-off if desired. This method is often used for threading a large number of small pieces.

Fig. 256. Die Mounted in Tailstock of Lathe for Threading Studs

A die may be mounted on lathe carriage for cutting screw threads, as shown in Fig. 257. The lead screw and half nuts are used to feed the die so that threads with a perfect lead are obtained.

Fig. 257. Die Mounted on Lathe Carriage for Cutting Accurate Screw Threads

Chapter XI

SPECIAL CLASSES OF WORK

There are many special classes of lathe work such as knurling, filing, polishing, coil winding, etc. The most important are illustrated and described briefly on the following pages.

Knurling

Knurling is the process of embossing the surface of a piece of work in the lathe with a knurling tool (Fig. 258) in the tool post of the lathe.

Three examples of knurling on a piece of steel are shown in Fig. 259. The pattern of the knurl is alike in all three cases but is of different grades, coarse, medium and fine.

Fig. 258. Knurling Tool for Lathe

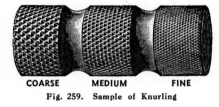

COARSE   MEDIUM   FINE

Fig. 259. Sample of Knurling

For all knurling operations the lathe should be arranged for the slowest back geared speed. After starting the lathe force the knurling tool slowly into the work at the right end until the knurl reaches a depth of about $\frac{1}{64}$ in. Then engage the longitudinal feed of the carriage and let the knurling tool feed across the face of the work. Plenty of oil should be used on the work during this operation.

When the left end of knurl roller has reached the end of work, reverse the lathe spindle and let the knurling tool feed back to the starting point. Do not remove the knurling tool from the impression but force it into the work another $\frac{1}{64}$ in., and let it feed back across the face of the work. Repeat this operation until the knurling is finished.

Fig. 260. Knurling a Steel Piece in the Lathe

87

Machining Work on the Face Plate

Before mounting a face plate on the spindle nose of the lathe, all dirt and chips should be removed from the threaded hole. Also clean the thread and the shoulder on the spindle nose because any dirt, chips or burrs will prevent the face plate from running true.

Oil the threads of the spindle so that the face plate will screw on easily and can be easily removed. If it seems difficult to screw on the face plate, unscrew the plate, remove the dirt, burrs, etc., and try again. The face plate hub should screw tight against the shoulder of the spindle but the face plate should not be spun up to the shoulder suddenly as this makes removal difficult.

The face plate is especially valuable in tool room work for machining holes in tools and jigs. In this class of work the holes must be accurately spaced, with an allowance usually not more than .001 of an inch.

Fig. 261. Boring an Eccentric Hole on the Face Plate of the Lathe

Clamping Work on Face Plate

Some care should be exercised when clamping work on the face plate so that neither the work nor the face plate will be sprung. A piece of paper placed between the face plate and the work will reduce the danger of the work slipping. Balance weights should be used as shown in Fig. 264.

Centering the Work

A center indicator may be used, as shown in Fig. 263, for accurately locating work on the face plate for drilling and boring. A dial indicator may also be used, as shown in Fig. 264.

Fig. 262. Boring a Bracket with an Angle Plate Attached to the Face Plate

Fig. 263. Centering Work on the Face Plate with a Center Indicator

Fig. 264. Locating Work on the Face Plate with a Dial Indicator

Filing and Polishing

All tool marks can be removed and a smooth, bright finish obtained on the surface of a piece of work by filing and polishing, as shown in Figs. 265 and 266.

Use a fine mill file and file with the lathe running at a speed so that the work will make two or three revolutions for each stroke of the file. File just enough to obtain a smooth surface. If too much filing is done the work will be uneven and inaccurate.

Fig. 265. Filing to Remove Tool Marks

Keep the left elbow high and the sleeves rolled up so there will be no danger from the lathe dog.

Keep the file clean and free from chips, using a file card frequently.

A very smooth, bright finish may be obtained by polishing with several grades of emery cloth after filing. Use oil on the emery cloth and run the lathe at high speed. Be careful not to let the emery cloth wrap around the revolving work.

Fig. 266. Polishing with Emery Cloth and Oil

Lapping

Hardened gauges, bushings and bearings are often finished in the lathe by lapping, as shown in Fig. 267. Emery cloth, emery dust and oil, diamond dust and other abrasives are used. Usually the lathe spindle is operated at high speed.

The lap may be very simple, consisting of a strip of emery cloth attached to a shaft, or it may be elaborately constructed of lead, copper, cast iron, etc. Some very fine and precise work may be accomplished by careful lapping.

Fig. 267. Lapping the Inside of a Hardened Steel Bushing with Emery Dust and Oil

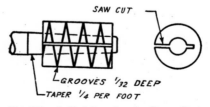

Fig. 268. A Cast Iron Lap for Emery Dust

Fig. 269. A Steel Lathe Mandrel

Machining Work on a Lathe Mandrel

Cylindrical work that has been bored and reamed in a chuck is usually further machined on a mandrel between the lathe centers, as shown in Figs. 270 and 271. The mandrel is slightly tapered and must be driven into the hole tight enough so that the work will not slip on the mandrel while it is being machined.

Large diameter work such as pulleys should be driven with a pin or driver attached to the lathe face plate if it can be arranged as this will eliminate possibility of the work slipping on the mandrel.

Before driving the mandrel into the hole in the work, oil both the mandrel and the hole so that the work will be easy to remove from the mandrel. If there is no lubricant on the mandrel it may "freeze" in the work, in which case it cannot be driven out without ruining both the work and the mandrel.

In driving a mandrel out of a piece of work be sure that it is driven in the opposite direction from that which it entered the work.

Standard lathe mandrels can be purchased in the various sizes. These mandrels are hardened and tempered and the surface that receives the work is ground usually to a taper of about .006 in. per foot.

In the case of special jobs having odd diameter holes, a soft mandrel may be used, turning and filing it to the proper diameter and tapered for a driving fit in the hole in the work.

Fig. 270. Turning a Pulley on a Mandrel

Fig. 271. Finishing a Bushing on a Mandrel

Special Mandrels

Special types of mandrels are often used for special classes of work. A nut mandrel for finishing the outside diameter of gear blanks is shown in Fig. 272. Expansion mandrels of various types are also available and are used where there is considerable variation in hole sizes.

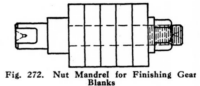

Fig. 272. Nut Mandrel for Finishing Gear Blanks

Winding Coils in the Lathe

The unusually wide range of positive power longitudinal feeds available on the lathe make it an ideal machine for winding electrical coils of all kinds. A revolution counter may be attached to register the number of turns, as shown in Fig. 273. Special gearing may be obtained for odd leads not in the usual thread cutting range of the lathe. Any type of coil form or wire guide required may be used.

Fig. 273. Winding a Coil

Spring Winding

Coil springs of all kinds may be wound on the lathe, as shown in Fig. 274. Special mandrels are used for irregular shaped springs. The lead screw and half nuts of the lathe are usually used to obtain a uniform lead so that the coils are all equally spaced.

Fig. 274. Winding a Spring

Boring Work Mounted on Lathe Carriage

Large work may be mounted on the lathe carriage for boring, as shown in Fig. 277.

The boring bar is held between centers and driven by a lathe dog. The work is clamped to the top of the lathe saddle and is fed to the tool by the automatic longitudinal feed of the carriage.

Several good types of boring bars for this class of work are shown in Figs. 275, 276, and 278.

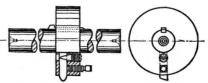

Fig. 275. Boring Bar with Fly Cutter

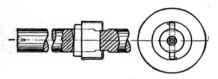

Fig. 276. Boring Bar with Boring Head

Fig. 277. Boring on the Lathe Carriage

Fig. 278. Boring Bar for Sizing the Hole

The Use of the Center Rest

The center rest is used for turning long shafts of small diameter and for boring and threading spindles. The end view of a center rest attached to the lathe bed is shown in Fig. 279.

To mount work in the center rest, first place the center rest on the lathe, then place the work between centers, slide the center rest to its proper position, and adjust the jaws upon the work. Careful adjustment is required because the work must revolve in these jaws. When the jaws are adjusted properly so that the work revolves freely, clamp the jaws in position, fasten the work to the head spindle of the lathe and slide the tailstock out of the way.

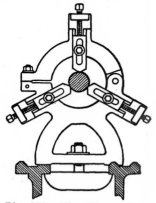

One end of the work may be held in a chuck, as shown in Fig. 280. For fine accurate work, the chuck should not be used.

Fig. 281 shows the method of fastening the work to the head spindle. The face plate is unscrewed from the shoulder about three or four turns. Then the work is tied securely to the face plate with several heavy belt laces, and the face plate is screwed onto the spindle. This tightens the lacing on the work and holds it firmly.

Fig. 279. The Center Rest
Mounted on the Lathe Bed

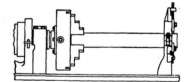

Fig. 280. Work Mounted in Chuck and
Center Rest

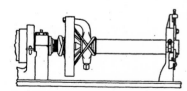

Fig. 281. Work Mounted on Center and
in Center Rest

Fig. 282. Cutting an Internal Screw Thread
with a Center Rest

Fig. 283. Supporting a Slender Shaft with
the Center Rest

The Use of the Follower Rest

The follower rest is attached to the saddle of the lathe to support work of small diameter that is liable to spring away from the cutting tool.

The adjustable jaws of the follower rest bear directly on the finished diameter of the work, as shown in Figs. 284 and 285. As the tool feeds along the work, the follower rest being attached to the saddle travels with the tool.

For the machining of small shafts in quantity, small rollers are sometimes substituted for the rigid adjustable jaws, and the device is then known as the Roller Bearing Follower Rest.

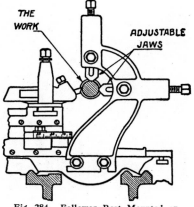

The application of both the center rest and follower rest at the same time is shown in Fig. 286. The spindles or shafts to be machined, while very small in diameter, are of considerable length, and in order to do a good job it is necessary to support the shaft with both the center rest and follower rest.

This method is used for machining small, delicate spindles used in textile mills.

Fig. 284. Follower Rest Mounted on Lathe Saddle

Fig. 285. Left—Threading a Long Slender Shaft with the Aid of a Follower Rest

Fig. 286. Right—Using Both the Center Rest and Follower Rest to Support a Long Slender Shaft

Fig. 287. No. 2-H South Bend Turret Lathe with Power Feed Bed Turret

Turret Lathes for Manufacturing

Turret Lathes are designed for the efficient production of duplicate parts. They are equipped with a power feed or hand lever feed turret having six faces, with automatic indexing and individual stop for each face. Cutting tools may be mounted in each of the six turret faces and indexed into position as required for performing various operations.

Turret lathes are usually equipped with either screw feed or hand lever operated double tool rest on the carriage cross slide. This permits using front and back tools for turning, facing, cutting off, and similar operations. A four-way turret tool post may be used on the cross slide.

Fig. 288. No. 1000 Series South Bend Turret Lathe with Hand Lever Operated Bed Turret

Fig. 289. Tooling on Turret Lathe

Fig. 290. Facing a Gear Blank

Fig. 291. Hand Lever Collet Chuck

Fig. 292. Hand Lever Tailstock

Fig. 293. Hand Lever Double Tool Rest

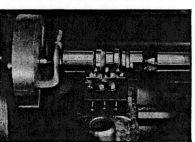

Fig. 294. Multiple Tool Block

Fig. 295. Two Tools Cutting Simultaneously on a Piece of Work

Fig. 296. An Irregular Shaped Piece Held in a Two-Jaw Chuck

Milling in the Lathe

The Milling and Keyway Cutting Attachment illustrated in Figs. 298 and 300 will take care of a great deal of milling in the small shop that does not have enough work to install an expensive milling machine.

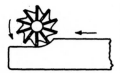

Fig. 297. Direction of Feed for Milling Operations

The cut is controlled by the hand wheel of the lathe carriage, the cross feed screw of the lathe and the vertical adjusting screw at the top of the milling attachment.

All milling cuts should be taken with the rotation of the cutter against the direction of the feed, shown in Fig. 297.

Fig. 298. Milling a Standard Keyway in a Shaft

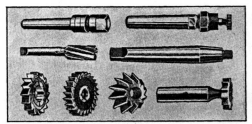

Fig. 299. An Assortment of Milling Cutters and Arbors

Fig. 300. Milling a Woodruff Keyway in a Shaft

Standard Keyways

The recognized standards for the depth and width of keyways in pulleys, gears, etc. are shown in Fig. 301 and the tabulation below. The same specifications are used for the depth and width of keyways in shafts.

The key should fit snugly in the keyway but must not be too tight.

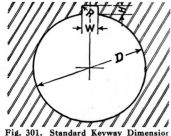

Fig. 301. Standard Keyway Dimension

Specifications of American Standard Keyways

Diameter Hole D Inches	Width W Inches	Depth H Inches	Radius R Inches	Diameter Hole D Inches	Width W Inches	Depth H Inches	Radius R Inches
1/2	3/32	3/64	.020	2 1/2	5/8	7/32	1/8
5/8 to 7/8	1/8	1/16	1/32	3	3/4	1/4	3/32
1	1/4	1/8	3/32	3 1/2	7/8	5/16	3/32
1 1/4	5/16	1/8	1/16	4	1	3/8	3/32
1 1/2	3/8	3/32	1/16	4 1/2	1 1/8	7/16	1/8
1 3/4	7/16	3/32	1/16	5	1 1/4	1/2	1/8
2	1/2	3/16	1/16				

Cutting Gears on the Lathe

The gear cutting attachment for the lathe, shown in Fig. 302, will cut spur and bevel gears of all kinds. It will do graduating and milling, external key seating, cutting at angles, splining, slotting and all regular dividing head milling work.

This attachment is practical for cutting small gears and for milling small light work of various kinds on the screw cutting lathe.

Fig. 302. Gear Cutting Attachment

The dividing head construction is based on the principle of interchangeable gears, the same as regularly used on gear cutting machines. The index plate shows the proper gears to use for divisions from 2 to 360.

Fig. 303. Cutting a Gear on a Lathe

Turning Wood, Fibre and Plastics

Turning wood in a metal working lathe is a very simple matter. Spur and cup centers are substituted for the 60° centers, a hand rest is attached and the lathe is ready for wood turning.

Special pulleys may be used on the motor and countershaft to provide a series of high spindle speeds for wood turning, in addition to the regular speeds for metal work.

Other materials may be machined as well. Alabaster, Catalin, Bakelite, fibre and other plastics, synthetic resins, etc., may be turned and polished with complete satisfaction.

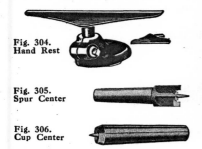

Fig. 304. Hand Rest

Fig. 305. Spur Center

Fig. 306. Cup Center

Fig. 307. Wood Turning in a Metal Working Lathe

Superfinished Bearing Surfaces

"Superfinish" is a term applied to a new process for producing a smooth surface on metals. The illustrations below show the comparative smoothness of a turned surface, a ground surface, and a superfinished surface. Superfinish can be applied to either a hardened surface or an unhardened surface.

| Turned Surface | Ground Surface | Superfinished Surface |

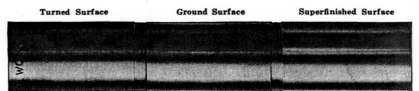

Fig. 308. Steel Shaft with Turned Surface, Ground Surface, and Superfinished Surface, actual size

 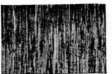 

| **A** | **B** | **C** |
| Turned Surface, Magnified 50 Diameters | Ground Surface, Magnified 50 Diameters | Superfinished Surface, Magnified 50 Diameters |

The process for producing a superfinished surface was developed by the Chrysler Corporation and is explained by them as follows:

"Superfinishing may be defined as an extremely fine crystalline surface finish produced upon flat, round, concave, convex and other types of surfaces, either external or internal. It is achieved by a combination of short motions, light abrasive pressure, slow abrasive cutting speeds, hard abrasive stones and a lubricant of proper viscosity to eliminate the amorphous scratches and surface defects created by previous mechanical operations, without causing new scratches and surface defects in the superfinished or crystalline surface".

When a metallic surface is finished by turning, the cutting action of the tool tears the metal, leaving a series of hills and valleys, as shown in photomicragraph "A". The crystalline structure of the metal is disturbed, leaving a fuzz of fragmented amorphous metal.

When a metallic surface is turned and then finished by grinding the cutting action of the grinding wheel is very much the same as the cutting action of the turning tool, and a similar series of hills and valleys are left, as shown in photomicrograph "B". This is because the surface of the grinding wheel is composed of a series of tiny cutting points which cut the metal in almost exactly the same way as the lathe tool. The crystalline structure of the metal is disturbed by the cutting action of the grinding wheel, leaving a fuzz of fragmented amorphous metal.

When either a turned or ground surface is superfinished, the fragmented amorphous metal is removed, and the surface produced is extremely smooth, as shown in the photomicrograph "C". A few minute scratches may be left, but most of the surface is microscopically smooth and makes an excellent bearing surface for a high speed shaft.

Micrometer Carriage Stop

The micrometer carriage stop consists of a micrometer spindle mounted in a clamp which may be securely locked onto the front V-way of the lathe bed, as shown in Fig. 310. A lock screw is provided for locking the spindle at any point.

The micrometer carriage stop is used for facing shoulders to an exact length. It is convenient for many production operations and is usually included in the equipment of all tool room lathes.

Fig. 310. Micrometer Carriage Stop

Open Side Tool Post

The open side or European tool post, shown in Fig. 311, is popular in Europe but is not used extensively in the United States.

This type of tool post holds the tool securely and is convenient for working close to the chuck or face plate. One advantage of this type of tool post is that the adjustment for the height of the cutting edge of the tool need not be lost when the tool is removed from the lathe if the entire tool post is removed with the tool. This permits changing tools more quickly than with the conventional type of tool post, provided several of the European tool posts are available.

Fig. 311. Open Side Tool Post

Metric Graduated Collars

Metric cross feed screw, metric compound rest screw and metric graduated collars are supplied on lathes that are to be used exclusively for working in the metric system.

The metric graduated collars read in tenths of a millimeter and are adjustable so that they may be set at zero whenever desired.

Fig. 312. Metric Graduated Collar

Metric Graduations on Taper Attachment

Taper attachments on lathes that are to be used for cutting tapers in the metric system are equipped with graduations reading in the metric system. Usually these graduations read in millimeters per centimeter and are in addition to the regular graduations.

Fig. 313. Taper Attachment with Metric Graduations

Metric Graduations on Tailstock Spindle

The tailstock spindle of the lathe may be graduated in centimeters, as shown in Fig. 314. The graduations are to aid in drilling accurately to the required depth.

Fig. 314. Metric Graduations on Tailstock Spindle

Grinding in the Lathe

When equipped with a good electric grinding attachment the lathe can be used for sharpening reamers and milling cutters, grinding hardened bushings and shafts and many other grinding operations.

The V-ways of the lathe bed should be covered with a heavy cloth or canvas to protect them from grit from the grinding wheel, and the lathe spindle bearings should also be protected. A small pan of water or oil placed just below the grinding wheel will collect most of the grit.

Fig. 315. External Grinding Attachment for the Lathe

A large, powerful grinder is most satisfactory for external grinding. The wheel should be at least four inches in diameter and the grinder should be mounted direct on the compound rest of the lathe, as shown in Fig. 315.

A small high speed grinder is best for internal grinding, as speed is more important than power for this class of work.

Fig. 316. Internal Grinding Attachment for Lathe

Grinding Wheel Speeds

The tabulation below shows grinding wheel speeds in revolutions per minute for surface speeds of 4000 and 5000 feet per minute.

Diam. Wheel	1 in.	2 in.	3 in.	4 in.	5 in.	6 in.	7 in.	8 in.	10 in.	12 in.
R.p.m. for surface Speed of 4,000 ft..............	15,279	7,639	5,093	3,820	3,056	2,546	2,183	1,910	1,529	1,273
R.p.m. for surface Speed of 5,000 ft..............	19,099	9,549	6,366	4,775	3,820	3,183	2,728	2,387	1,910	1,592

GRINDING WHEELS FOR VARIOUS KINDS OF WORK

Tabulation shows grade of Norton Grinding wheels.

Kind of Work	Rough Grind	Finish Grind
Cast Iron................	3736-K Crystolon	3760-J Crystolon
Soft Steel..............	46-M5BE	60-M5BE
Hardened Steel.........	3846-L5BE	3860-L5BE
High Speed Steel........	3846-K5BE	3860-K5BE
Brass or Bronze........	3736-K Crystolon	3760-J Crystolon
General Work...........	46-N5BE	46-N5BE
Aluminum	30-M3L Shellac	36-M3L Shellac
Bakelite	3736-K Crystolon	3746-K Crystolon
Soft Rubber............	3720-K5T-2 Crystolon Bakelite	3746-K5T-2 Crystolon Bakelite
Hard Rubber............	3730-K5T-2 Crystolon Bakelite	3760-K5T-2 Crystolon Bakelite
Automobile Valves......	1960-M	80-L6BE
Tungsten Carbide........	3760/1-17 Crystolon	37100/2-H7 Crystolon

Diamond Dresser for Truing Grinding Wheel

The grinding wheel must be balanced and must be dressed with a diamond dresser if a smooth, accurate ground finish is to be obtained. The grinding wheel must be dressed frequently as it is used to keep it true and free from particles of metal which become embedded in the periphery of the wheel.

The diamond dresser consists of a small industrial diamond mounted in a steel shank, as shown in Fig. 318. The dresser must be rigidly supported in a fixture for truing the grinding wheel, as shown in Fig. 317.

Fig. 317. Truing a Grinding Wheel with a Diamond Dresser

The diamond point of the dresser should be placed on center, or slightly below center and the revolving grinding wheel passed back and forth across the diamond. Remove about .001 in. from the wheel at each cut and dress the wheel just enough to make it run true.

Fig. 318. Diamond Dresser

Grinding Hardened Steel Parts

Hardened steel parts should be carefully ground in order to produce a smooth, accurate finish. The part should be machined to within a few thousandths of the finished size before it is hardened. After hardening, all scale should be removed before grinding. Remove only a few thousandths at each pass of the grinding wheel for if the part is ground too fast it may become overheated and warp, or the temper may be drawn.

Fig. 319. Grinding a Hardened Steel Bushing

Sharpening Reamers and Cutters

Reamers and milling cutters may be sharpened by grinding in the lathe, as shown in Figs. 320, 321 and 322. Some reamers are first circular ground, then relieved by grinding with a tooth rest set slightly below center, as shown in Fig. 320, leaving a land .002 in. to .005 in. wide. Other reamers and most milling cutters are ground with about 2° relief.

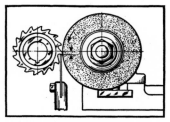

Fig. 320. Grinding Clearance on a Milling Cutter

Fig. 321. Grinding an Angular Cutter in the Lathe

Fig. 322. Grinding a Straight Reamer in the Lathe

Fig. 323. A 10-in. Swing Bench Lathe with 1-in. Collet Capacity

1-in. Collet Capacity Lathe

The 10-in. Swing, 1-in. Collet Capacity Lathe is used for tool room and manufacturing operations on parts made from bar and tubing stock. This type of lathe has a wide range of spindle speeds so that both small diameter parts and large diameter parts can be machined efficiently.

The headstock spindle is unusually large for a lathe of this size, having a 1⅜ in. hole. Work up to 1⅜ in. in diameter can be passed through the spindle and held in the 3-jaw universal chuck for machining.

Large Collet Capacity

The unusually large collet capacity is obtained by using a spindle with an extra large hole. Both hand wheel type draw-in collet chuck attachment and hand lever type draw-in collet chuck attachment are used. The hand wheel type attachment is usually selected for tool room use and the hand lever type attachment is preferred for most manufacturing operations.

Fig. 324. Making Shoulder Screws from 1-in. Bar Stock Held in the Collet

Modern
Shops
Equipped with
South Bend
Precision
Lathes.

Fig. 327. Truing and Undercutting an Armature Commutator in the Lathe

The Lathe in the Auto Service Shop

The Back-Geared, Screw Cutting Lathe is frequently called the "Universal Tool," and this applies in automotive service work as well as in general industry. Most of the mechanical parts of the automobile, bus, truck, tractor and airplane are originally made on lathes or in special machines which are adaptations of the lathe.

A lathe with 9-in. or 10-in. swing is very practical for handling such jobs as refacing valves; truing armature commutators and undercutting mica; finishing pistons; beveling piston skirts; reaming piston pin holes; making bushings, bearings and glands; boring rebabbitted connecting rods, and many other jobs. Special attachments used on the lathe greatly increase its versatility.

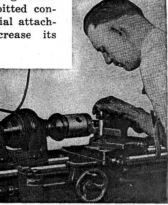

Fig. 328. Making a Replacement Bushing Complete in a 9-In. Lathe

Fig. 329. Turning a Semi-Machined Piston to Size in a 9-In. Lathe

Armature Truing

Machining the commutator of an armature true and under-cutting the mica are two of the most important jobs in auto electrical work, and these jobs are most easily handled in the lathe.

A small lathe equipped for these jobs is shown in Fig. 330. The undercutting attachment is mounted on the lathe in such a way that it is ready for instant use, yet it does not interfere with turning the commutator.

Fig. 330. Undercutting an Armature Commutator in the Lathe

Refacing Valves

A lathe equipped with a grinding attachment and a special hollow valve chuck for refacing valves is shown in Fig. 331.

Other valve jobs done in the lathe include: Truing the valve tappet face and rocker arm face; making valve guide bushings and valve seat replacement rings, etc.

Fig. 331. Refacing a Valve by Grinding in the Lathe

Finishing Pistons

Pistons of all sizes and types can be rough and finish turned in the lathe, as shown in Fig. 332. The lathe can also be used for reaming and honing piston pin holes, cutting oil grooves in pistons, remachining piston ring grooves, beveling piston skirts, etc.

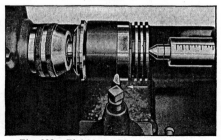

Fig. 332. Finishing a Piston in the Lathe

Boring Connecting Rods

A lathe equipped with a connecting rod boring attachment shown in Fig. 333 is the most practical machine for boring rebabbitted connecting rods. Rods of all sizes can be tested for alignment, rebored, faced and finish trimmed.

Fig. 333. Boring a Rebabbitted Connecting Rod in the Lathe

Machining Eccentrics

A simple eccentric can be machined on a straight mandrel having two sets of center holes as shown in Fig. 334. One set of centers is used for machining the concentric hub and the other set of centers is used for machining the eccentric part.

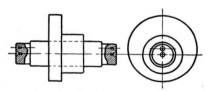

Fig. 334. A Mandrel with Two Sets of Center Holes for Machining an Eccentric

Crankshaft Turning

Crankshaft turning is an adaptation of eccentric machining. A single throw crankshaft mounted in the lathe for machining the throw bearing is shown in Fig. 335. The adapters attached to each end of the crankshaft have offset center holes corresponding to the throw of the crankshaft.

Fig. 335. A Crankshaft Mounted in the Lathe for Machining the Throw Bearing

Truing Crankshaft Bearings

Throw bearings of automobile crankshafts are often worn out of round or scored and must be remachined. A special crank pin truing tool, shown in Fig. 336, permits truing the throw bearings without the use of offset centers. The tool travels around with the throw bearing and is so constructed that it will machine the bearing round and straight. The lathe spindle must revolve very slowly (about 10 r.p.m.) while this tool is being used.

Fig. 336. Truing the Throw Bearings of a Crankshaft

Testing Crankshafts

Crankshafts may be tested between the lathe centers, as shown in Fig. 337. The dial indicator mounted in the tool post of the lathe reads in thousandths of an inch and will show exactly how much the crankshaft is sprung and will also disclose any out-of-round condition of the bearing. Straightening a crankshaft is a delicate job and should be attempted only by an experienced mechanic.

Fig. 337. Testing a Crankshaft in the Lathe

Fig. 338. A Portable Machine Shop Built in a Large Truck

Portable Machine Shop

The portable machine shop shown in Fig. 338 is rapidly gaining in popularity. This type of shop is especially valuable for service in oil fields, construction camps, air ports, army posts, etc., also for the maintenance of road building equipment and for repairing construction machinery and equipment on large engineering projects. The advantage in taking the shop to the job is obvious when the delay and difficulty involved in transporting heavy, awkward parts to and from the shop are taken into consideration.

The equipment of the portable machine shop may be quite complete, consisting of a 16 in. by 8 ft. lathe, a 20-in. drill press, a forge, anvil, grinder, welding outfit, etc., as shown above, or it may be limited to a small lathe and a good assortment of small tools. The equipment will vary with the purpose of the shop and the amount to be invested.

The lathe is the most important of all of the tools in the portable machine shop because it can be used for so many classes of work. When equipped with the necessary attachments, the lathe may be used as a milling machine, gear cutting machine, grinding machine, drill press, etc.

Regardless of the size of the lathe and other equipment, it is important that the truck or trailer in which the shop is installed be of substantial construction with a good solid floor. Provision should be made for blocking and leveling the floor while the machinery is in use. All parts must be securely fastened in place so there will be no danger of damage while the shop is being transported from one location to another.

Power for operating the lathe and other machinery is usually obtained through a generator installed in such a way that it can be operated by the truck engine. This same generator also supplies current for electric lights to illuminate the shop and also large flood lights which permit working near the unit after night.

Fig. 339. Metal Turning on a Small Back-Geared Screw Cutting Lathe

Metal Work Done on a Small Lathe

A surprising variety of work can be done on a small back-geared screw cutting precision lathe. The metal parts shown below are examples of precision work machined on a 9-in. Back-Geared Screw Cutting Bench Lathe. Operations handled include screw thread cutting, turning, boring, reaming, drilling, filing, polishing, milling, etc.

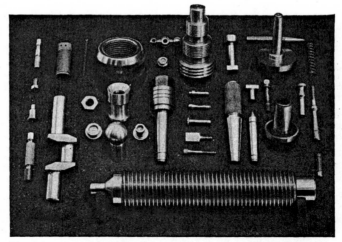

Fig. 340. Metal Parts Machined Entirely on a 9-in. Swing Back-Geared Screw Cutting Precision Lathe

Fig. 341. Model "A" South Bend 9 in. Swing Precision Lathe

THE SMALL METAL WORKING LATHE

The new Model "A" South Bend 9 in. Precision Lathe shown above is well adapted for work of small diameter requiring a high degree of accuracy and sensitivity. This lathe has a maximum collet capacity of ½ in., takes work up to 9⅛ in. in diameter over the bed ways, and will swing 5½ in. over the saddle cross slide.

Twelve spindle speeds ranging from 41 to 1270 r.p.m. are provided by the underneath motor drive which is enclosed in the cabinet under the lathe headstock. Smooth operation at high speed is attained by using a direct belt drive to the spindle. A wrenchless bull gear lock permits engaging the back gears quickly when slow speeds are required.

Practical attachments for this lathe include handwheel or hand lever type draw-in collect chuck attachment, double tool rest, hand lever bed turret attachment, taper attachment, micrometer carriage stop, thread dial indicator, oil pump and piping.

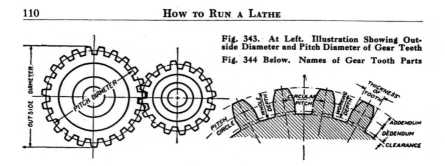

Fig. 343. At Left. Illustration Showing Outside Diameter and Pitch Diameter of Gear Teeth

Fig. 344 Below. Names of Gear Tooth Parts

Information on Gears

The rules and formulas listed below may be used for calculating the dimensions of involute spur gears.

Diametral Pitch—Number of teeth divided by pitch diameter, or 3.1416 divided by circular pitch.

Example: If a gear has 40 teeth and the pitch diameter is 4 in., the diametral pitch is 40 divided by 4, and the diametral pitch is 10, or in other words there are 10 teeth to each inch of the pitch diameter, and the gear is 10 diametral pitch.

Circular Pitch—Distance from center to center of two adjacent teeth along pitch circle, or 3.1416 divided by diametral pitch.

Pitch Diameter—Number of teeth divided by diametral pitch.

Example: If the number of teeth is 40 and the diametral pitch is 4, divide 40 by 4, and the quotient, 10, is the pitch diameter.

Outside Diameter—Number of teeth plus two divided by diametral pitch.

Example: If the number of teeth is 40 and the diametral pitch is 4, add 2 to the 40, making 42, and divide by 4; the quotient, 10½, is the outside diameter of gear or blank.

Addendum—1 divided by diametral pitch.

Whole Depth of Tooth—2.157 divided by diametral pitch.

Thickness of Tooth—1.5708 divided by diametral pitch.

Number of Teeth—Pitch diameter multiplied by diametral pitch, or multiply outside diameter by diametral pitch and subtract 2.

Example: If the diameter of the pitch circle is 10 in. and the diametral pitch is 4, multiply 10 by 4 and the product, 40, will be the number of teeth in the gear.

Example: If the outside diameter is 10½ and the diametral pitch is 4, multiply 10½ by 4 and the product, 42, less 2, or 40, is the number of teeth.

Center Distance—Total number of teeth in both gears divided by two times the diametral pitch.

Example: If the two gears have 50 and 30 teeth respectively, and are 5 pitch, add 50 and 30, making 80, divide by 2, and then divide the quotient, 40, by the diametral pitch, 5, and the result, 8 in., is the center distance.

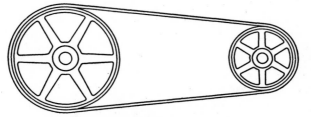

Fig. 345. A Pair of Pulleys for Flat Belt Drive

Calculating the Speed and Size of Pulleys

Diameter of Driving Pulley—Multiply the diameter of the driven pulley by its number of revolutions, and divide by the number of revolutions of the driver.

Diameter of Driven Pulley—Multiply the diameter of the driving pulley by its number of revolutions, and divide the product by the number of revolutions of the driven pulley.

Speed of Driven Pulley—Multiply the diameter of the driving pulley by its number of revolutions, and divide by the diameter of the driven pulley.

Speed of Driving Pulley—Multiply the diameter of the driven pulley by its number of revolutions, and divide by the diameter of the driving pulley.

The driving pulley is called the driver and the driven pulley is the driven or follower.

R.P.M. indicates the number of revolutions per minute.

Example: Problem 1.

Given: Speed of the driving pulley 260 R.P.M. Speed of the driven pulley 390 R.P.M. Diameter of the driven pulley 8 in.

To find the diameter of the driving pulley.

$$390 \times 8 = 3120$$
$$3120 \div 260 = 12$$

The diameter of the driving pulley is 12 in.

Width of Pulleys—Pulleys for flat belts should be about 10% wider than the width of the belt used.

Types of Pulleys—Two types of pulleys are used for flat belts, the crowned face pulley and the flat face pulley. Crowned pulleys should always be used if possible, as it is the crown that keeps the belt on the pulley. Flat face pulleys should be used only when it is necessary to shift the belt from one position on the pulley to another, as in a drum pulley or a wide faced pulley on a machine used to match a tight and loose pulley on a countershaft.

Fitting a Chuck Plate to a Chuck

Before a chuck can be used on a lathe it must be fitted with a chuck plate that has been threaded to fit the spindle nose of the lathe. Semi-machined chuck plates that have been accurately threaded to fit the lathe spindle can be obtained from the lathe manufacturer.

Mounting Chuck Plate on Spindle

Before screwing the chuck plate on to the spindle nose of the lathe, clean the threads of the chuck plate and the spindle nose thoroughly. Make sure that there are no chips, burrs or small particles of dirt lodged in the screw threads, or on face of hub, and also make sure that the shoulder on the headstock spindle is perfectly clean and free from chips or burrs.

Oil the threads of the headstock spindle and the chuck plate and screw the chuck plate onto the spindle nose. Do not jam the threads tight or it may be difficult to remove the chuck plate after it has been finished.

Fig. 346. Semi-Machined Chuck Plate, Threaded to Fit the Spindle Nose of the Lathe

Finishing the Flange

First, machine the face of the chuck plate, taking one roughing cut about $\frac{1}{32}$ in. deep, and then one or two finishing cuts, removing not over .001 in. in the last cut.

Measure the diameter of the recess in the back of the chuck carefully with inside calipers, and set outside calipers to correspond with the inside calipers. Machine the diameter of the chuck plate flange very carefully. Take very light finishing cuts and try the chuck on the chuck plate frequently, as the chuck plate must fit snugly into the recess in the back of the chuck.

Fig. 347. Rear View of the Lathe Chuck

After the chuck plate has been finished to fit the recess in the back of the chuck, remove it from the lathe spindle and chalk the face of the flange thoroughly. Place the flange in the recess in the back of the chuck and tap lightly on the chuck plate so that the edge of the bolt holes in the chuck will mark the location of bolt holes.

Drill the holes $\frac{1}{16}$ in. larger in diameter than the bolts used to secure the chuck plate onto the chuck. It is very important that the bolt holes be large enough to eliminate all possibility of the bolts binding.

Fig. 348. Chuck with Chuck Plate Attached

Hardening and Tempering Lathe Tools

After a forged lathe tool has been used for some time, it should be re-forged, hardened and tempered. If carefully done, this will make the tool as good as new. Before attempting to harden and temper the tool, make sure of the kind of steel from which it is made.

To Distinguish Carbon Tool Steel

To distinguish carbon tool steel from high speed steel, touch against emery wheel. Carbon steel gives off a shower of bright yellow sparks; high speed steel gives off a few dark red sparks.

Fig. 349. Hardening a Lathe Tool

To Harden Carbon Tool Steel

To harden a forged lathe tool made of carbon tool steel, heat the end of the tool slowly to a bright cherry red for a distance of at least an inch back of the cutting edge; then immerse the point about 1½ in. deep in cold water, but do not cool the shank. When the point is cool remove from water, polish the cutting edge with emery cloth and wipe with oily rag.

The tool is now hardened, and as the heat in the shank passes into the point it will discolor the polished surface, indicating the amount the temper is drawn. When a light straw color appears, cool the entire tool quickly in water and it will have the correct strength and toughness for metal turning.

The method outlined above may be followed for hardening and tempering any tool made of carbon tool steel. For wood cutting tools, taps and dies, draw to a dark straw color. For hatchets, screw drivers, cold chisels, etc., draw to a brown yellow; for springs dark purple.

Case Hardening

To case harden a piece of machinery steel, heat the steel to a cherry red; then remove from fire and apply cyanide of potassium to the surface you wish to case harden. The cyanide will dissolve slowly and be absorbed by the steel. After the surface has received a thorough coat of cyanide return the steel to the fire and heat slowly for about one minute so that the cyanide will be thoroughly absorbed by the steel. Remove from the fire and quench in cold water.

How to Anneal Tool Steel

Carbon tool steel may be annealed by heating slowly and evenly to a cherry red and then placing in a box of lime or ashes to cool slowly. The steel should be completely covered and when it is cooled to room temperature will be ready for machining.

How to Anneal Brass

Brass that has been hardened through cold working may be annealed by heating to a dull red when held in dark shadow and plunging into cold water. Care must be taken not to overheat the brass.

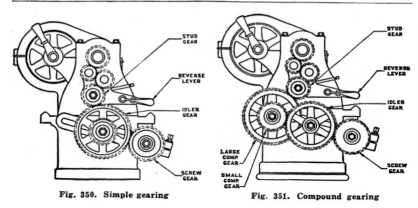

Fig. 350. Simple gearing   Fig. 351. Compound gearing

How to Calculate Change Gears for Thread Cutting

If it is necessary to cut a special thread that does not appear on the index chart of a lathe or if no index chart is available, the gears required can easily be calculated. All South Bend Lathes are even geared; that is, the stud gear revolves the same number of revolutions as the headstock spindle, and when gears of the same size are used on both the lead screw and stud, the lead screw and spindle revolve the same number of revolutions, so it is not necessary to consider the gearing between the headstock spindle and the stud gear when calculating change gears.

If simple gearing is to be used, as shown in Fig. 350, the ratio of the number of teeth in the change gears used will be the same as the ratio between the thread to be cut and the thread on the lead screw. For example, if 10 threads per inch are to be cut on a lathe having a lead screw with 6 threads per inch, the ratio of the change gears would be 6 to 10. These numbers may be multiplied by any common multiplier to obtain the number of teeth in the change gears that should be used.

Rule—To calculate change gears, multiply the number of threads per inch to be cut and the number of threads per inch in the lead screw by the same number.

Example: Problem—To cut 10 threads per inch on lathe having lead screw with 6 threads per inch.
Solution— 6 x 8 = 48 — No. of teeth in gear on stud.
10 x 8 = 80 — No. of teeth in gear on lead screw.

If these gears are not to be found in the change gear set, any other number may be used as a common multiplier, such as 3, 5, 7, etc.

When compound gearing, as shown in Fig. 351, is used, the ratio of the compound idler gears must also be taken into consideration, but otherwise the calculations are the same as for simple gearing. Usually, the compound idler gear ratio is 2 to 1, so that the threads cut are just twice the number per inch as when simple gearing is used.

DECIMAL EQUIVALENTS OF FRACTIONAL PARTS OF AN INCH

$\frac{1}{64}$= .015625	$\frac{11}{32}$ = .34375	$\frac{11}{16}$ ...... .6875
$\frac{1}{32}$ = .03125	$\frac{23}{64}$= .359375	$\frac{45}{64}$= .703125
$\frac{3}{64}$= .046875	$\frac{3}{8}$ .... .375	$\frac{23}{32}$ = .71875
$\frac{1}{16}$ ..... .0625	$\frac{25}{64}$= .390625	$\frac{47}{64}$= .734375
$\frac{5}{64}$= .078125	$\frac{13}{32}$ = .40625	$\frac{3}{4}$ .... .75
$\frac{3}{32}$ = .09375	$\frac{27}{64}$= .421875	$\frac{49}{64}$= .765625
$\frac{7}{64}$= .109375	$\frac{7}{16}$ ..... .4375	$\frac{25}{32}$ = .78125
$\frac{1}{8}$ ...... .125	$\frac{29}{64}$= .453125	$\frac{51}{64}$= .796875
$\frac{9}{64}$= .140625	$\frac{15}{32}$ = .46875	$\frac{13}{16}$ ..... .8125
$\frac{5}{32}$ = .15625	$\frac{31}{64}$= .484375	$\frac{53}{64}$= .828125
$\frac{11}{64}$= .171875	$\frac{1}{2}$ 5	$\frac{27}{32}$ = .84375
$\frac{3}{16}$ ..... .1875	$\frac{33}{64}$= .515625	$\frac{55}{64}$= .859375
$\frac{13}{64}$= .203125	$\frac{17}{32}$ = .53125	$\frac{7}{8}$ .... .875
$\frac{7}{32}$ = .21875	$\frac{35}{64}$= .546875	$\frac{57}{64}$= .890625
$\frac{15}{64}$= .234375	$\frac{9}{16}$ ..... .5625	$\frac{29}{32}$ = .90625
$\frac{1}{4}$ .... .25	$\frac{37}{64}$= .578125	$\frac{59}{64}$= .921875
$\frac{17}{64}$= .265625	$\frac{19}{32}$ = .59375	$\frac{15}{16}$ ..... .9375
$\frac{9}{32}$ = .28125	$\frac{39}{64}$= .609375	$\frac{61}{64}$= .953125
$\frac{19}{64}$= .296875	$\frac{5}{8}$ .... .625	$\frac{31}{32}$ = .96875
$\frac{5}{16}$ ..... .3125	$\frac{41}{64}$= .640625	$\frac{63}{64}$= .984375
$\frac{21}{64}$= .328125	$\frac{21}{32}$ = .65625	1 .... 1.
	$\frac{43}{64}$= .671875	

METRIC AND ENGLISH LINEAR MEASURE

The measuring rules shown below are graduated, in the Metric system and in the English system. This shows at a glance the comparison of the fractions of the Metric and English units, the meter and the inch.

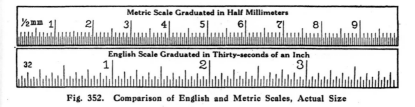

Fig. 352. Comparison of English and Metric Scales, Actual Size

TABLE OF METRIC LINEAR MEASURE

10 Millimeters	= 1 Centimeter	1 Centimeter	=	.3937 inch
10 Centimeters	= 1 Decimeter	1 Decimeter	=	3.937 inches
10 Decimeters	= 1 Meter	1 Meter	=	39.37 inches

Equivalents of Millimeters in Decimals of Inches

$\frac{1}{10}$ mm = .00394 in.	8 mm = .31496 in.	18 mm = .70866 in.
$\frac{1}{5}$ mm = .00787 in.	9 mm = .35433 in.	19 mm = .74803 in.
$\frac{1}{2}$ mm = .01969 in.	10 mm = .39370 in.	20 mm = .78740 in.
1 mm = .03937 in.	11 mm = .43307 in.	21 mm = .82677 in.
2 mm = .07874 in.	12 mm = .47244 in.	22 mm = .86614 in.
3 mm = .11811 in.	13 mm = .51181 in.	23 mm = .90551 in.
4 mm = .15748 in.	14 mm = .55118 in.	24 mm = .94488 in.
5 mm = .19685 in.	15 mm = .59055 in.	25 mm = .98425 in.
6 mm = .23622 in.	16 mm = .62992 in.	26 mm = 1.02362 in.
7 mm = .27559 in.	17 mm = .66929 in.	27 mm = 1.06299 in.

SHOP KINKS

From "American Machinist"

A good lubricant for turning, boring and milling aluminum is made of equal parts of lard oil and kerosene. Kerosene alone is also good, and cheap.

When drilling or turning hard steel (not hardened steel) in the lathe, run slowly and lubricate the tool with turpentine, or turpentine and spirits of camphor.

Red lead and graphite are good lubricants for the tail center. On heavy work, make the countersinks for the center as large as possible without making the job unsightly.

Probably the handiest lathe chuck for the jobbing shop is the four-jawed independent chuck having stepped, reversible jaws. These will hold almost any shaped piece, and hold it firmly.

Before complaining that the lathe is not in line, be sure that the bed is carefully leveled up. Don't twist the lathe bed out of shape by pulling it down on to an uneven floor with lagscrews, and then expect it to turn straight.

Squared ends and a uniform depth of lathe center are desirable for accurate and economical work. Don't make the mistake of thinking that any boy can do the centering well enough without proper instruction and supervision.

A little square of mica with a tin rim makes a good chip guard when turning brass. A simple wire spring clip can be fastened to the tin rim so that the guard can be readily fastened to or detached from the tool or tool post.

If a $\frac{1}{32}$-in. slot is made on the top of the tail center running from the point back to a little beyond the large part of the conical end, the center can be oiled without slacking it back. This is better than grinding half the conical end away.

The indicator, of the dial variety, is an extremely handy tool around the engine lathe, particularly if accurate work is to be done. Don't get the idea that this is an unnecessary frill, but get accustomed to using it if you wish to become an accurate workman.

Universal lathe chucks are very convenient to have in the shop but can rarely be depended on to run true if accuracy is desired. They may be plenty good enough for a large proportion of the work to be done, but accurate work demands separate adjustment of the jaws.

The lathe centers should be well cared for, the point ground in place, if possible, and always put into place in the same position. This may sound finicky but it is the only way to produce accurate work. Don't make the mistake of using a lathe center either for a hammer or a center punch.

DON'TS FOR MACHINISTS

From "Machinery"

Don't run a lathe with the belt too loose.

Don't run the point of your lathe tool into the mandrel.

Don't rap the chips out of your file on the lathe shears.

Don't set a lathe tool below the center for external work.

Don't start up a lathe without seeing that the tail stock spindle is locked.

Don't put an arbor or shaft on lathe centers without lubricant on them.

Don't leave too much stock on a piece of work to take off with the finishing cut.

Don't try a steel gauge or an expensive caliper on a shaft while it is running.

Don't put a mandrel into a newly bored hole without a lubricant of some kind on it.

Don't put a piece of work on centers unless you know that the internal centers are clean.

Don't try to straighten a shaft on lathe centers, and expect that the centers will run true afterwards.

Don't put a piece of work on lathe centers unless you know that all your centers are at the same angles.

Don't take a lathe center out of its socket without having a witness mark on it, and put it back again according to the mark.

Don't start polishing a shaft on lathe centers without having it loose enough to allow for the expansion by heat from the polishing process.

Don't run your lathe tool into the faceplate.

Don't try to knurl a piece of work without oiling it.

Don't run a lathe an instant after the center begins to squeal.

Don't forget to oil your machine every morning; it works better.

Don't forget that a fairly good center-punch may be made from a piece of round file.

Don't forget that a surface polished with oil will keep clean much longer than one polished dry.

Don't start to turn up a job on lathe centers unless you know that the centers are both in line with the ways.

Don't cross your belt laces on the side next the pulley, for that makes them cut themselves in two.

Don't try to cut threads on steel or wrought iron dry; use lard oil or a cutting compound.

Don't run a chuck or faceplate up to the shoulder suddenly; it strains the spindle and threads and makes removal difficult.

Don't screw a tool post screw any tighter than is absolutely necessary; many mechanics have a false idea as to how tight a lathe tool should be to do its work.

To drive the center out of head spindle use a rod and drive through the hole in spindle.

When putting a lathe chuck on the head spindle, always remove the center.

When the center is removed from the head spindle of the lathe, always put a piece of rag in spindle hole to prevent any dirt from collecting.

Fig. 353.   A Trade School Shop, equipped with 38 South Bend Lathes

INDUSTRIAL APPRENTICE TRAINING
In the United States

The industrial plants of the United States are very much interested in the Vocational and Trade Schools.   These schools are doing remarkable work for the individual boys of their own communities and also for the industries of the entire United States as a whole.

The metal-working industry is aware that well equipped vocational schools can teach young men the fundamentals of the machinist's trade much better than they can be taught to apprentices in factories. Under the guidance of a capable and thoroughly trained instructor, the boys not only receive practical instruction in the operation of various machine tools, but are also taught the necessary shop mathematics, mechanical drawing, business English, economics, etc.

Industry looks to vocational and trades training schools to supply young men with sufficient vocational training so that they can be further trained in the factory for positions of expert workmen, specialty mechanics, foremen, superintendents, salesmen and advertising men—not for the ordinary jobs of operating production machines.

A number of large industrial plants in the United States have established training courses in their own plants to supplement and continue the fundamental training now being given in the Vocational and Trade Schools.

Fig. 354. A High School Shop, equipped with 43 South Bend
Precision Lathes

Fig. 355. Another High School Shop, equipped with 17 South Bend
Precision Bench Lathes

MOTORS FOR OPERATING LATHES

Many different types of motors are available for operating individual motor driven lathes. In some cases any one of several different motors may be used, but usually there is one particular type of motor that will give more satisfactory service than any other. A brief study of the various types will aid in selecting the most desirable motor.

Reversing Motor Required—Several important lathe operations necessitate reversing the rotation of the lathe spindle. For this reason it is important that the motor be of the reversible type. The motor should conform exactly with the electric current that is to be used. Dual rated motors may be operated on two different voltages.

When more than one type of current is available, the motor should be selected for use with the most desirable type of current. Two or three-phase alternating current is generally considered more desirable for power service than single phase current or direct current.

The Size of Motor specified by the lathe manufacturer should always be used unless the lathe is to be operated at spindle speeds higher than standard. More power is required for high speed operation, necessitating a larger motor. The lathe manufacturer should always be consulted before equipping a lathe for high speed operation. The speed of the motor should be, as nearly as possible, the speed specified.

Polyphase Motors for 3-phase or 2-phase alternating current are instantly reversible, have a high starting torque and constant speed. These motors have no brushes or commutators to cause radio interference. They are very satisfactory for lathe operation.

Capacitor Motors for single phase alternating current are made in instant reversing, and start-stop reversing types. The instant reversing type is preferable but the start-stop reversing type is less expensive and may be used satisfactorily on certain classes of lathe work. This type of motor has a fairly high starting torque and constant speed. It has a starting switch but has no commutator to cause radio interference.

Repulsion Induction Motors for single phase alternating current have a fair starting torque and fairly constant speed. They are made to order in the instant reversing type for lathe operation. Having a commutator and brushes, this motor may cause radio interference.

Split Phase Motors for single phase alternating current do not have sufficient starting torque for satisfactory use with a lathe.

Shunt Wound Motors for direct current are instantly reversible, have a constant torque and a very constant speed. They are entirely satisfactory for lathe operation, although the commutator and brushes may cause radio interference.

Compound Wound Motors for direct current are instantly reversible. They have a high starting torque and a fairly constant speed. The compound wound motors are satisfactory for lathe operation but may cause radio interference due to the commutator and brush construction.

Universal Motors are designed for operating on either alternating current or direct current. This type of motor is unsatisfactory for operating a metal working lathe.

REFERENCE BOOKS ON LATHE WORK

The booklets listed below are 6 x 9 in. or 8½ x 11 in. and contain from 12 to 32 pages each. When ordering specify titles. Money order or stamps of any country accepted. All prices are postpaid.

"Modern School Shops" Booklet No. 55-R. Illustrates some of the most modern school shops in the United States. 24 pages 8½" x 11"..................................**Sent Free**

"The South Bend Machine Shop Course for Apprentice Training". Detail drawings and operation sequence sheets for twelve practical machine shop projects. 32 pages 8½" x 11" **50c**

Bulletin H-1—"Keep Your Lathe Clean". Shows how protecting the lathe from abrasive dirt will increase production, reduce scrap, and lengthen the life of the lathe .... **Sent Free**

Bulletin H-2—"Oiling the Lathe". Explains the importance of adequate lubrication..................**Sent Free**

Bulletin H-3—"Leveling the Lathe". Gives detailed information on the correct installation and leveling of the lathe for precision machine work..........**Sent Free**

"How to Cut Screw Threads in the Lathe" Bulletin No. 36-A. Complete information on cutting screw threads. 24 pages 6" x 9"...**10c**

"How to Grind Lathe Tool Cutter Bits" Bulletin No. 35. Shows how to grind and sharpen lathe tool cutter bits. 16 pages 6" x 9"..**10c**

"How to Run a Lathe" in Spanish, Portuguese, and French

The book "How to Run a Lathe" has been translated into the Spanish, Portuguese, and French languages. The translated editions are complete and are practically identical with the English edition. They are the same size and contain the same number of pages and illustrations.

All translations have been prepared by competent engineers who are familiar with the mechanical terms used in the various languages.

A copy of "How to Run a Lathe" in the Spanish, Portuguese, or French language will be mailed anywhere in the world, postpaid for 25c, stamps or money order of any country accepted. State language wanted.

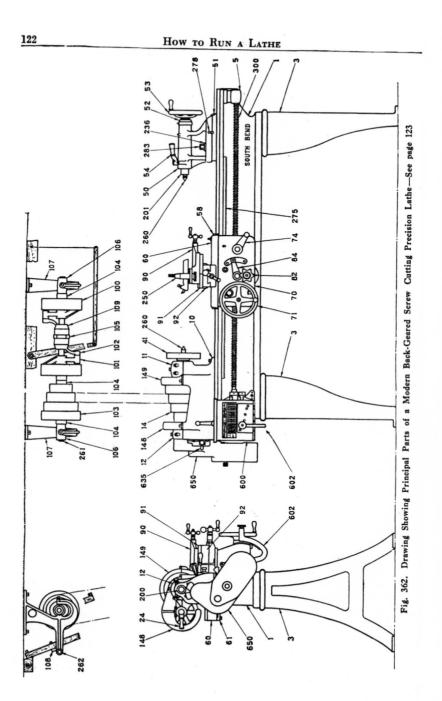

Fig. 362. Drawing Showing Principal Parts of a Modern Back-Geared Screw Cutting Precision Lathe—See page 123

NAMES AND NUMBERS OF LATHE PARTS

The number and the name of the principal parts of the lathe are tabulated on this and the following page. Parts marked thus (*) are shown in Figs. 362 and 363. The name of any numbered part shown in Figs. 362 and 363 will be found listed below opposite the corresponding number.

Part No.	Name of Part	Part No.	Name of Part
*1	Bed	73	Lead Screw Half Nut Gib
*3	Long Legs	*74	Nut Cam
4	Lead Screw Bracket, Front	74A	Nut Cam Lever
*5	Lead Screw Bracket, Rear	75	Nut Cam Friction Washer
*10	Headstock	76	Rack Pinion Gear
*11	Headstock Cap, Large	77	Apron Worm Wheel
11A	Headstock Cap Shims, Large	77B	Worm Wheel Lock Ring
*12	Headstock Cap, Small	78	Oil Distributing Washer
12A	Headstock Cap Shims, Small	79	Worm Bushing
13	Headstock Clamp (2)	80	Apron Clutch Sleeve
*14	Spindle Cone	81	Apron Clutch Plate Nut
15	Bull Gear	81A	Apron Clutch Discs (Inner)
16	Bull Gear Clamp	81B	Apron Clutch Discs (Outer)
17	Cone Pinion	*82	Apron Clutch Knob
18	Quill Gear	83	Idler Gear Shifter
19	Quill Sleeve	*84	Idler Gear Shifter Lever
20	Quill Sleeve Pinion	85	Apron C. F. Gear
21	Ecc. Shaft Bushing, Rear	86	Apron Idler Gear
21F	Ecc. Shaft Bushing, Front	87	Apron C. F. Pinion
22	Bronze Box, Large	88	Idler Gear Shifter Knob
23	Bronze Box, Small	89	Idler Gear Shifter Knob Plunger
*24	Back Gear Lever	89A	Idler Gear Shifter Knob Spring
25	Spindle Take-up Nut		
25A	Spindle Take-up Nut Screw	89B	Hand Wheel Pinion Thrust Spring
26	Spindle Take-up Nut Washer		
27	Reverse Twin Gears (2)	*90	Compound Rest Top
28	Reverse Gear	*91	Compound Rest Swivel
30	Spindle Reverse Gear	*92	Compound Rest Base
39	Thrust Collar on Lead Screw	*94	Compound Rest Bushing
40	Large Face Plate	*95	Compound Rest Nut
*41	Small Face Plate	*96	Compound Rest Chip Guard
*50	Tailstock Top	97	Apron Feed Lock Plunger
*51	Tailstock Base	98	Apron Feed Lock Plunger Arm
*52	Tailstock Nut	*100	C. S. Clutch Pulleys (2)
*53	Tailstock Hand Wheel	*101	C. S. Clutch Spiders (2)
*54	Tailstock Binding Lever	*102	C. S. Clutch Levers (2)
55	Tailstock and Change Gear Wrench	*103	C. S. Cone
56	Tailstock Clamp	103W	C. S. Cone Counterweight
*58	Saddle Felt Retainer	*104	C. S. Collars (4)
59	Saddle Felt	*105	C. S. Yoke Lever
*60	Saddle	*106	C. S. Boxes (2)
*61	Saddle Gib	*107	C. S. Hangers (2)
62	Saddle Lock	*108	C. S. Shipper Nut
63	Cross Feed Bushing	*109	C. S. Yoke Cone
64	Cross Feed Gra. Collar	*110	Comp. Rest Graduated Collar
*65	Cross Feed Nut	*148	Quill Gear Guard
*66	Cross Feed Nut Shoulder Screw	*149	Bull Gear Guard
67	Thread Cutting Stop	154	Thread Cutting Index Chart
*70	Apron	*200	Headstock Spindle
*71	Apron Hand Wheel	*201	Tailstock Spindle
72	Lead Screw Half Nut	202	Back Gear Eccentric Shaft

Part No.	Name of Part	Part No.	Name of Part
203	Apron Worm	249	T. S. Binding Plug, Lower
204	Apron Rack Pinion	*250	Tool Post
205	Spindle Sleeve	*251	Tool Post Ring
206	Tailstock Binding Lever Screw	*252	Tool Post Wedge
207	Spindle Thrust Collar	253	Tool Post Wrench
207A	Ball Thrust Bearing for Spindle	255	C. S. Clutch Set Screw Wrench
208	Apron Worm Collar	*257	Compound Rest Top Tapered Gib
*209	Tool Post Block	*260	Centers (2)
210	Carriage Lock Collar Screw	*261	C. S. Shaft
*211	Compound Rest Clamp Bolts	*262	C. S. Shipper Rod
*211A	Compound Rest Lock Pin	263	C. S. Expansion Cam
214	Bull Gear Clamp Plunger	264	C. S. Shipper Nut Washer
215	Apron Clutch Sleeve Hex. Nut	271	Apron Hand Wheel Handle
*219	Dovetail Gib Adjusting Screw	272	Half Nut Cam Lever Handle
*219A	Gib Adjusting Screw Binder Plug	273	Tailstock Hand Wheel Handle
219B	Gib Adjusting Screw Binder	*275	Rack
*221	Compound Rest Base Tapered Gib	276	Cross Feed Ball Crank
223	Automatic Apron Clutch Screw	*277	Compound Rest Ball Crank
224	Cross Feed Screw	*278	Tailstock Set-Over Screws (2)
'225	Apron Hand Wheel Pinion	279	Tailstock Clamp Bolt
226	Tailstock Screw	282	Headstock Oiler
227	Apron Worm Wheel Trough	*283	Tailstock Clamp Nut
227A	Apron Worm Wheel Trough Gasket	284	Reverse Bracket Oiler
229	Twin Gear Studs (2)	286	Tailstock Spindle Key
*230	Compound Rest Screw	287	Lead Screw Bracket Oiler
231	Auto. Cross Feed Stud	289	Oil Hole Plug
232	Apron Half Nut Studs (2)	291	Q. C. G. Box Hub Oiler Tube
*233	Tool Post Screw	292	T. S. Oil Reservoir Plug
*236	Tailstock Nut Washer	*295	Nut-C. F. Screw
237	Reverse Shaft Nut	296	Nut-C. R. Screw
239	Apron Worm Key	*300	Lead Screw
243	C. S. Ball Point Set Screws (2)	*600	Gear Box
244	Fillester Head Screws Apron to Saddle (4)	*602	Gear Box Tumbler
245	Thread Cutting Stop Thumb Screw	617	Top Lever of Gear Box
246	Back Gear Lug Set Screws (2)	*635	Reverse Bracket
248	T. S. Binding Plug, Upper	636	Reverse Shaft Key
		637	Reverse Shaft
		638	Reverse Spring Latch
		*650	Primary Gear Guard
		662	Sliding Gear Plate
		664	Size of Lathe Plate

PRINCIPAL PARTS OF COMPOUND REST

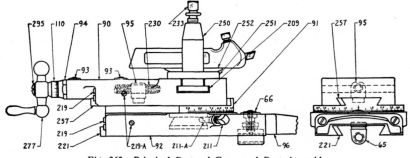

Fig. 363. Principal Parts of Compound Rest Assembly

The Surface Plate

The surface plate, shown in the illustration, is a flat cast iron plate used in the building of fine machinery to test plane surfaces while hand-scraping. Three surface plates are necessary so that they may be tested together occasionally and the surface kept perfectly true and flat by rescraping when it becomes worn.

Fig. 364. Toolmaker Locating Holes in Jig Plate with Aid of Surface Plate, Surface Gauge, and Gauge Blocks

Surface plates are also used by toolmakers for locating and checking holes in jig plates, laying out work, testing, inspecting, and similar operations. With the aid of a surface gauge and a set of gauge blocks very accurate work can be done on the surface plate.

Hand-Scraping on Lathe Bed

After a lathe bed has been machined, it is thoroughly seasoned, then finish planed. Extreme accuracy is obtained by scraping the ways by hand, so all lathe beds are hand-finished and frosted by master craftsmen during the process of fitting the carriage, headstock, and tailstock.

The base of the headstock and tailstock are hand-scraped to fit perfectly onto the V-ways of the lathe bed. The V-ways of the saddle are also scraped to conform with the V-ways of the lathe bed.

Fig. 365. Hand-Scraping the V-ways of a Lathe Bed

General Lathe Catalog

CATALOG OF SOUTH BEND LATHES
Mailed Free to Any Address on Request

This new catalog describes the entire line of South Bend Engine Lathes, Toolroom Lathes, Turret Lathes and attachments. The Engine Lathes and Toolroom Lathes range in size from 9" to 16" swing. The Turret Lathes are made in three sizes having 9", 10", and 16" swings.

A copy of this catalog will be mailed on request, postpaid, no charge. Specify size of lathe and type of work when requesting catalog.

Photo Courtesy of J. Walter Thompson Co.

MOTION PICTURES ON LATHE WORK

To speed up the training of lathe operators for national defense industries, the South Bend Lathe Works has sponsored the production of two 16 mm sound motion pictures in full color. Professionally filmed, these pictures show practical shop methods as practiced in modern industrial plants. Showing time for each of the two 800 ft. reels is approximately 20 minutes.

The first reel entitled "The Lathe" clearly shows what a lathe is, what a lathe is for, and how the various parts operate. Important operations, including turning, facing, and thread cutting, are demonstrated. The second reel "Plain Turning" shows in detail each operation in the machining of a cylindrical shaft between the lathe centers. Close-ups show locating and drilling of center holes, application of cutting tools, use of calipers and micrometers, and operation of the lathe carriage and apron.

Bookings for recognized organizations teaching machine shop practice are being scheduled on a free loan basis. Write now for application blank to insure an early showing. These films are also available for outright purchase.

INDEX

Subject	Page
Accuracy of a Screw Cutting Lathe	40
Acme Screw Threads	83
Alignment of Centers	48, 51
American National Screw Thread	70, 71
Annealing Steel and Brass	113
Application of Lathe Tools	34
Apprentice Training	118
Apron of the Lathe	14, 24
Armature Truing and Undercutting	104, 105
Attachments for the Lathe	57, 58, 62, 81, 84, 92 to 109
Automatic Carriage Feeds	24, 25, 49, 114
Auto Service Shop Lathe Work	104
Back-Geared Headstock	12, 22
Bearings for Headstock Spindle	13, 125
Bearing Surfaces, Superfinished	98
Bed of Lathe	13, 125
Belts, Lacing, Shifting, etc.	17, 19, 26, 111
Belt Tension	19
Bench Lathes	4, 10, 94, 102, 108, 109
Bits for Lathe Tool Holders	27 to 35
Books for the Mechanic	121
Boring in the Lathe	33, 42, 56, 59, 63, 91, 92
Brown & Sharpe Worm Thread	83
Calipers, Use of	37 to 39
Capacity of the Lathe	11
Carriage of Lathe	14, 24
Carriage Stop, Micrometer	99
Case Hardening	113
Catalog of South Bend Lathes	125
Center Gauge	60, 75
Center Holes, Locating and Drilling	43 to 46
Center Indicator	54, 88
Center Rest	92
Centers, Alignment of	48, 51
Centers, Mounting in Lathe Spindle	47
Centers, Removing from Lathe Spindle	47
Centers, Truing	60
Centering Work in Chuck	54, 55
Centering Work on Face Plate	88
Change Gear Chart	73
Change Gears, Calculating	114
Chuck Plate, Fitting to Chuck	112
Chuck Work	53
Chucks, Mounting on Lathe Spindle	54
Chucks, Practical Sizes	55
Chucks, Removing from Lathe Spindle	55
Coil Winding in the Lathe	91
Collet Chuck and Collets	57, 58, 95, 102
Combination Center Drill and Countersink	45
Commutator Truing and Undercutting	104, 105
Compound Gearing for Threads and Feeds	72, 114
Compound Rest of Lathe	59, 60, 77, 78
Connecting Rod Boring	105
Countershaft Drive Lathe	18
Countershaft Speeds	23
Crankshaft Testing	106
Crankshaft Turning	106
Cross Feed of Lathe, Automatic	25, 49
Crotch Center	65
Cup Center	97
Cutter Bits	27 to 35
Cutting Power of Lathe	36
Cutting Screw Threads	69 to 86
Cutting Speeds for Various Metals	50

Subject	Page
Decimal Equivalents	115
Dial Test Indicator	55, 88
Die for Cutting Screw Threads	86
Direct Cone Drive for Lathe Spindle	22
Don'ts for Machinists	117
Draw-in Collet Chuck Attachment	57, 102
Drilling in the Lathe	65
Drilling Center Holes	45, 46
Drilling Cored Hole	67
Drill Chuck	56
Drill Pad	65
Drills, How to Sharpen	67
Eccentric Machining	106
Emery Wheel Speeds	100
Face Plate Work	88
Facing Work on Centers	49
Filing in the Lathe	89
Fitting Chuck Plate to Chuck	112
Floating Reamer Driver	68
Follower Rest for Lathe	93
Forged Steel Lathe Tools	35, 113
Gear Box, Operation of	74
Gear Cutting in the Lathe	97
Geared Screw Feed Lathe	4
Gearing Lathe for Cutting Screw Threads	72
Gears, Information on	110
Grinding Cutter Bits	29
Grinding in the Lathe	100
Grinding Wheel Speeds	100
Grinding Wheels, Truing	101
Hand Lever Double Tool Rest	95
Hand Lever Draw-in Collet Chuck	58, 95
Hand Lever Tailstock	95
Hand Rest for Wood Turning	97
Hand-Scraping Bearing Surfaces	125
Hand Wheel Draw-in Collet Chuck	57
Hardening and Tempering	113
Headstock of Lathe	12, 22, 125
Headstock Spindle Chuck	56
Height of Cutting Edge of Cutter Bit	28, 32, 33
High Speed Steel Cutter Bits	27 to 34
History of the Lathe	3
Horizontal Motor Drive for Bench Lathes	10
Independent Chuck	54
Indicator, Center and Dial	54, 55
Industrial Apprentice Training	118
Keyways, American Standard Sizes	96
Knurling in the Lathe	87
Lacing Belts	17
Lapping in the Lathe	89
Large Collet Capacity Lathe	102
Lathe Dog, Attaching	47, 48
Lathe Tools	27
Left Hand Thread	80
Leveling the Lathe	15, 16
Locating Center Holes	43
Longitudinal Feed, Automatic	25, 49, 114
Machine Shop Course for Schools	121
Mandrels, Machining Work on	90
Manufacturing, Turret Lathes for	94
Measurements, How to Take Accurate	37

INDEX

Subject	Page
Measuring Screw Threads	79
Metric Graduations	99
Metric Lathes	85
Metric Measure	115
Metric Micrometer	39
Metric Screw Threads	84
Micrometer Carriage Stop	99
Micrometer Calipers, How to Read	39
Micrometer Collar	24, 78
Milling in the Lathe	96
Morse Standard Tapers	64
Motion Pictures on Lathe Work	126
Motor Drives for Lathes	8, 9, 10
Motors, Hints on Selecting Correct Types	120
Mounting Chuck on Lathe Spindle	54
Mounting Lathe Center in Spindle	47
Multiple Screw Threads	86
Names of Lathe Parts	21, 123
Notes on Belts and Pulleys	17, 19, 26, 111
Notes on Gears	110
Notes on Lathe Work	26
Nut Mandrel	90
Oiling the Lathe	20
One Inch Collet Lathe	102
Open Side Tool Post	35, 99
Operation of Lathe	21
Part Numbers of Lathe Parts	123
Pedestal Motor Drive for Lathe	8
Piston Finishing	105
Pitch and Lead of Screw Thread	70
Plain Turning	43
Polishing in the Lathe	89
Portable Machine Shop	107
Power Carriage Feeds	25, 49
Precision Level	16
Production Attachments	95
Pulleys, Calculating Speed and Size	26, 111
Quick Change Gear Box for Threads and Feeds	74
Quick Change Gear Lathes	6, 25, 74
Reamer Sharpening	101
Reaming in the Lathe	68
Refacing Automobile Valves	105
Reference Books on Lathe Work	121
Removing Chuck from Spindle	55
Reverse Lever on Headstock	22
Roughing Cuts, Maximum Depth	36
Rough Turning, Cutter Bit for	30
School Shops	118, 119
Scraping on Lathe Bed	125
Screw Pitch Gauge	79
Screw Threads, calculating change gears for	114
Screw Threads, Fitting and Testing	79
Screw Thread Tables	71
Screw Thread Terms	70
Screw Thread Cutting	69
Selecting a Lathe	11

Subject	Page
Semi-machined Chuck Plate, Fitting	112
Setting Cutter Bit for Threads	76
Setting the Lathe Tool	28, 49
Setting Over Tailstock for Taper Turning	61
Shifting Belts	19
Shop Kinks	116
Shoulders, Measuring and Machining	52
Size and Capacity of Lathe	11
Special Classes of Work	87, 125
Spindle Speeds of Lathes	23, 50
Spring Winding in the Lathe	91
Spur Center	97
Square Screw Threads	82
Standard Change Gear Lathe	5, 25, 72
Standard Screw Thread Tables	71
Step Chuck and Closer	58
Stellite Cutter Bits	35
Superfinished Bearing Surfaces	98
Surface Plate	125
Tailstock Adjustment	51
Tailstock of Lathe	24
Tailstock Set-over for Tapers, Calculating	61
Tap Drill Sizes	71
Taper Attachment	62
Taper Gauges	61
Taper Turning and Boring	59-63
Tapered Screw Threads	82
Tapers, Standard Dimensions of	64
Tapping in the Lathe	68
Tempering and Hardening	113
Test for Alignment of Lathe Centers	48, 51
Test for Taper Fit	61
Testing Instruments for Chuck Work	54, 55
Testing the Lathe for Accuracy	16, 40
Thread Cutting	69
Thread Cutting Stop	78
Thread Cutting Tool	32, 75
Thread Dial	81
Thread Tool Gauge	75
Threads, Terms Relating to	70
Tool Bits, High Speed Steel	27 to 34
Tool Holders for Lathes	27
Tool Room Lathes	7
Transposing Gears for Metric Screw Threads	84
Tungsten Carbide Cutter Bits	35
Turning Between Centers	43
Turret for Lathe Bed	95
Turret Lathes for Manufacturing	94
Undercutting Armature Commutators	105
Underneath Motor Drive for Lathe	9, 109
Universal Chuck	56
U. S. Standard Screw Thread	70
Valve Refacing in the Lathe	105
Whitworth Screw Thread	83
Wood Turning on Lathe	97
Worm Thread, 29°	83

How to Become a Machinist

1. Keep your cutting tools sharp.
2. Look at your drawing carefully before starting your job.
3. Be sure your machine is set up right before starting the work.
4. Take your measurements accurately.
5. Keep your machine well oiled, clean and neat. Personal neatness will give you personality.
6. Take an interest in your job; don't feel that you are forced to work.
7. Learn the fundamentals of mechanical drawing.
8. Keep your belts tight and free from oil.
9. Take as heavy a cut as the machine and cutting tool will stand until you are near the finished size; then finish carefully and accurately.
10. Try to understand the mechanism of the machine you are operating.
11. Hold yourself responsible for the job you are working on.
12. Keep your eyes on the man ahead of you; you may be called on to take his place some day.
13. Have a place for everything, and keep everything in its place.
14. Read one or two of the technical magazines relating to your line of work.
15. If a boy learns a trade properly he becomes a first-class mechanic, but if he has ability he need not stop at that.

    Henry Ford, George Westinghouse and others got their start because they were mechanics.
16. If you have spoiled a job, admit your carelessness to your foreman, and don't offer any excuses.
17. Before starting to work on a lathe, roll up your sleeves and remove your necktie—safety pays.

SOUTH BEND LATHE WORKS

NOTE: A Blue Print (13 in. x 22 in.) of the above seventeen suggestions, suitable for wall display, will be mailed upon the receipt of 10c to cover cost of mailing.

CPSIA information can be obtained at www.ICGtesting.com
Printed in the USA
LVOW12s0548070315

429478LV00005B/438/P